TRACTATUS LOGICO-PHILOSOPHICUS

International Library of Philosophy and Scientific Method

EDITOR: TED HONDERICH

For list of books in the series see back end paper

TRACTATUS

LOGICO-PHILOSOPHICUS

The German text of

Ludwig Wittgenstein's

Logisch-philosophische Abhandlung

with a new edition of the Translation by
D. F. Pears & B. F. McGuinness

and with the Introduction by
BERTRAND RUSSELL, F.R.S.

LONDON
ROUTLEDGE & KEGAN PAUL
NEW YORK: THE HUMANITIES PRESS

First German edition in Annalen der Naturphilosophie, 1921
First English edition, with a translation, 1922
Second impression, with a few corrections, 1933
& many subsequent impressions.
This translation first published 1961
by Routledge *&* Kegan *Paul Ltd*
Broadway House, 68–74 *Carter Lane*
London, E.C.4
Second impression, with a few corrections, 1963
Third impression 1966
Fourth impression 1969
Second edition of this translation 1971
Printed in Great Britain
by Richard *Clay (The Chaucer Press), Ltd*
Bungay, Suffolk

ISBN 0 7100 3004 5

TRANSLATORS' PREFACE 1960

In the German text of the present edition small errors have been remedied and a greater measure of consistency and clarity in punctuation and spelling has been sought. All previous editions and impressions have been consulted. The English translation is new, and not a revision of the earlier translation. The index of English words has been designed to serve also as a guide to the German terminology.

The translators are grateful to Lord Russell for permission to reprint his introduction to the edition of 1922. They have not altered the translations it contains, which are those of Lord Russell himself or of the first English translator.

The translators wish to thank Miss G. E. M. Anscombe, Professor Max Black, Mr Burton Dreben, Professor Erich Heller, Professor Gilbert Ryle, and many Oxford colleagues, for valuable suggestions and criticisms. They themselves are responsible for the errors that remain.

TRANSLATORS' PREFACE 1971

IN THIS NEW EDITION the German text has been corrected at a few points, and a fair number of small changes have been made in the translation. Most of the latter were made in the light of Wittgenstein's own suggestions and comments in his correspondence with C. K. Ogden about the first translation. This correspondence has now been published by Professor G. H. von Wright (Blackwell, Oxford, 1971).

CONTENTS

INTRODUCTION

BY BERTRAND RUSSELL, F.R.S.

MR WITTGENSTEIN'S *Tractatus Logico-Philosophicus*, whether or not it prove to give the ultimate truth on the matters with which it deals, certainly deserves, by its breadth and scope and profundity, to be considered an important event in the philosophical world. Starting from the principles of Symbolism and the relations which are necessary between words and things in any language, it applies the result of this inquiry to various departments of traditional philosophy, showing in each case how traditional philosophy and traditional solutions arise out of ignorance of the principles of Symbolism and out of misuse of language.

The logical structure of propositions and the nature of logical inference are first dealt with. Thence we pass successively to Theory of Knowledge, Principles of Physics, Ethics, and finally the Mystical (*das Mystische*).

In order to understand Mr Wittgenstein's book, it is necessary to realize what is the problem with which he is concerned. In the part of his theory which deals with Symbolism he is concerned with the conditions which would have to be fulfilled by a logically perfect language. There are various problems as regards language. First, there is the problem what actually occurs in our minds when we use language with the intention of meaning something by it; this problem belongs to psychology. Secondly, there is the problem as to what is the relation subsisting between thoughts, words, or sentences, and that which they refer to or mean; this problem belongs to epistemology. Thirdly, there is the problem of using sentences so as to convey truth rather than falsehood; this belongs to the special sciences dealing with the subject-matter of the sentences in question. Fourthly, there is the question: what relation must one fact (such as a sentence) have to another in order to be *capable* of being a symbol for that other? This last is a logical question, and is the one with which Mr Wittgenstein is concerned.

He is concerned with the conditions for *accurate* Symbolism, i.e. for Symbolism in which a sentence 'means' something quite definite. In practice, language is always more or less vague, so that what we assert is never quite precise. Thus, logic has two problems to deal with in regard to Symbolism: (1) the conditions for sense rather than nonsense in combinations of symbols; (2) the conditions for uniqueness of meaning or reference in symbols or combinations of symbols. A logically perfect language has rules of syntax which prevent nonsense, and has single symbols which always have a definite and unique meaning. Mr Wittgenstein is concerned with the conditions for a logically perfect language—not that any language is logically perfect, or that we believe ourselves capable, here and now, of constructing a logically perfect language, but that the whole function of language is to have meaning, and it only fulfils this function in proportion as it approaches to the ideal language which we postulate.

The essential business of language is to assert or deny facts. Given the syntax of a language, the meaning of a sentence is determinate as soon as the meaning of the component words is known. In order that a certain sentence should assert a certain fact there must, however the language may be constructed, be something in common between the structure of the sentence and the structure of the fact. This is perhaps the most fundamental thesis of Mr Wittgenstein's theory. That which has to be in common between the sentence and the fact cannot, so he contends, be itself in turn *said* in language. It can, in his phraseology, only be *shown*, not said, for whatever we may say will still need to have the same structure.

The first requisite of an ideal language would be that there should be one name for every simple, and never the same name for two different simples. A name is a simple symbol in the sense that it has no parts which are themselves symbols. In a logically perfect language nothing that is not simple will have a simple symbol. The symbol for the whole will be a 'complex', containing the symbols for the parts. In speaking of a 'complex' we are, as will appear later, sinning against the rules of philosophical grammar, but this is unavoidable at the outset. 'Most propositions and questions that have been written about philosophical matters are not false but senseless. We cannot, therefore, answer questions of this kind at all, but only state their senselessness. Most questions and propositions of the philosophers result from the fact that we do not

understand the logic of our language. They are of the same kind as the question whether the Good is more or less identical than the Beautiful' (4.003). What is complex in the world is a fact. Facts which are not compounded of other facts are what Mr Wittgenstein calls *Sachverhalte*, whereas a fact which may consist of two or more facts is called a *Tatsache*: thus, for example, 'Socrates is wise' is a *Sachverhalt*, as well as a *Tatsache*, whereas 'Socrates is wise and Plato is his pupil' is a *Tatsache* but not a *Sachverhalt*.

He compares linguistic expression to projection in geometry. A geometrical figure may be projected in many ways: each of these ways corresponds to a different language, but the projective properties of the original figure remain unchanged whichever of these ways may be adopted. These projective properties correspond to that which in his theory the proposition and the fact must have in common, if the proposition is to assert the fact.

In certain elementary ways this is, of course, obvious. It is impossible, for example, to make a statement about two men (assuming for the moment that the men may be treated as simples), without employing two names, and if you are going to assert a relation between the two men it will be necessary that the sentence in which you make the assertion shall establish a relation between the two names. If we say 'Plato loves Socrates', the word 'loves' which occurs between the word 'Plato' and the word 'Socrates' establishes a certain relation between these two words, and it is owing to this fact that our sentence is able to assert a relation between the persons named by the words 'Plato' and 'Socrates'. 'We must not say, the complex sign "*a*R*b*" says "*a* stands in a certain relation R to *b*"; but we must say, *that* "*a*" stands in a certain relation to "*b*" says *that a*R*b*' (3.1432).

Mr Wittgenstein begins his theory of Symbolism with the statement (2.1): 'We make to ourselves pictures of facts.' A picture, he says, is a model of the reality, and to the objects in the reality correspond the elements of the picture: the picture itself is a fact. The fact that things have a certain relation to each other is represented by the fact that in the picture its elements have a certain relation to one another. 'In the picture and the pictured there must be something identical in order that the one can be a picture of the other at all. What the picture must have in common with reality in order to be able to represent it after its manner—rightly or falsely—is its form of representation' (2.161, 2.17).

We speak of a logical picture of a reality when we wish to imply only so much resemblance as is essential to its being a picture in any sense, that is to say, when we wish to imply no more than identity of logical form. The logical picture of a fact, he says, is a *Gedanke*. A picture can correspond or not correspond with the fact and be accordingly true or false, but in both cases it shares the logical form with the fact. The sense in which he speaks of pictures is illustrated by his statement: 'The gramophone record, the musical thought, the score, the waves of sound, all stand to one another in that pictorial internal relation which holds between language and the world. To all of them the logical structure is common. (Like the two youths, their two horses and their lilies in the story. They are all in a certain sense one)' (4.014). The possibility of a proposition representing a fact rests upon the fact that in it objects are represented by signs. The so-called logical 'constants' are not represented by signs, but are themselves present in the proposition as in the fact. The proposition and the fact must exhibit the same logical 'manifold', and this cannot be itself represented since it has to be in common between the fact and the picture. Mr Wittgenstein maintains that everything properly philosophical belongs to what can only be shown, to what is in common between a fact and its logical picture. It results from this view that nothing correct can be said in philosophy. Every philosophical proposition is bad grammar, and the best that we can hope to achieve by philosophical discussion is to lead people to see that philosophical discussion is a mistake. 'Philosophy is not one of the natural sciences. (The word "philosophy" must mean something which stands above or below, but not beside the natural sciences.) The object of philosophy is the logical clarification of thoughts. Philosophy is not a theory but an activity. A philosophical work consists essentially of elucidations. The result of philosophy is not a number of "philosophical propositions", but to make propositions clear. Philosophy should make clear and delimit sharply the thoughts which otherwise are, as it were, opaque and blurred' (4.111 and 4.112). In accordance with this principle the things that have to be said in leading the reader to understand Mr Wittgenstein's theory are all of them things which that theory itself condemns as meaningless. With this proviso we will endeavour to convey the picture of the world which seems to underlie his system.

The world consists of facts: facts cannot strictly speaking be defined, but we can explain what we mean by saying that facts are what make propositions true, or false. Facts may contain parts which are facts or may contain no such parts; for example: 'Socrates was a wise Athenian', consists of the two facts, 'Socrates was wise', and 'Socrates was an Athenian'. A fact which has no parts that are facts is called by Mr Wittgenstein a *Sachverhalt*. This is the same thing that he calls an atomic fact. An atomic fact, although it contains no parts that are facts, nevertheless does contain parts. If we may regard 'Socrates is wise' as an atomic fact we perceive that it contains the constituents 'Socrates' and 'wise'. If an atomic fact is analysed as fully as possible (theoretical, not practical possibility is meant) the constituents finally reached may be called 'simples' or 'objects'. It is not contended by Wittgenstein that we can actually isolate the simple or have empirical knowledge of it. It is a logical necessity demanded by theory, like an electron. His ground for maintaining that there must be simples is that every complex presupposes a fact. It is not necessarily assumed that the complexity of facts is finite; even if every fact consisted of an infinite number of atomic facts and if every atomic fact consisted of an infinite number of objects there would still be objects and atomic facts (4.2211). The assertion that there is a certain complex reduces to the assertion that its constituents are related in a certain way, which is the assertion of a *fact*: thus if we give a name to the complex the name only has meaning in virtue of the truth of a certain proposition, namely the proposition asserting the relatedness of the constituents of the complex. Thus the naming of complexes presupposes propositions, while propositions presuppose the naming of simples. In this way the naming of simples is shown to be what is logically first in logic.

The world is fully described if all atomic facts are known, together with the fact that these are all of them. The world is not described by merely naming all the objects in it; it is necessary also to know the atomic facts of which these objects are constituents. Given this totality of atomic facts, every true proposition, however complex, can theoretically be inferred. A proposition (true or false) asserting an atomic fact is called an atomic proposition. All atomic propositions are logically independent of each other. No atomic proposition implies any other or is inconsistent with any other. Thus the whole business of logical inference is concerned

with propositions which are not atomic. Such propositions may be called molecular.

Wittgenstein's theory of molecular propositions turns upon his theory of the construction of truth-functions.

A truth-function of a proposition *p* is a proposition containing *p* and such that its truth or falsehood depends only upon the truth or falsehood of *p*, and similarly a truth-function of several propositions *p*, *q*, *r*, ... is one containing *p*, *q*, *r*, ... and such that its truth or falsehood depends only upon the truth or falsehood of *p*, *q*, *r*, ... It might seem at first sight as though there were other functions of propositions besides truth-functions; such, for example, would be '*A* believes *p*', for in general *A* will believe some true propositions and some false ones: unless he is an exceptionally gifted individual, we cannot infer that *p* is true from the fact that he believes it or that *p* is false from the fact that he does not believe it. Other apparent exceptions would be such as '*p* is a very complex proposition' or '*p* is a proposition about Socrates'. Mr Wittgenstein maintains, however, for reasons which will appear presently, that such exceptions are only apparent, and that every function of a proposition is really a truth-function. It follows that if we can define truth-functions generally, we can obtain a general definition of all propositions in terms of the original set of atomic propositions. This Wittgenstein proceeds to do.

It has been shown by Dr Sheffer (*Trans. Am. Math. Soc.*, Vol. XIV. pp. 481–488) that all truth-functions of a given set of propositions can be constructed out of either of the two functions 'not-*p* or not-*q*' or 'not-*p* and not-*q*'. Wittgenstein makes use of the latter, assuming a knowledge of Dr Sheffer's work. The manner in which other truth-functions are constructed out of 'not-*p* and not-*q*' is easy to see. 'Not-*p* and not-*p*' is equivalent to 'not-*p*', hence we obtain a definition of negation in terms of our primitive function: hence we can define '*p* or *q*', since this is the negation of 'not-*p* and not-*q*', i.e. of our primitive function. The development of other truth-functions out of 'not-*p*' and '*p* or *q*' is given in detail at the beginning of *Principia Mathematica*. This gives all that is wanted when the propositions which are arguments to our truth-function are given by enumeration. Wittgenstein, however, by a very interesting analysis succeeds in extending the process to general propositions, i.e. to cases where the propositions which are arguments to our truth-function are not given by enumeration

but are given as all those satisfying some condition. For example, let fx be a propositional function (i.e. a function whose values are propositions), such as 'x is human'—then the various values of fx form a set of propositions. We may extend the idea 'not-p and not-q' so as to apply to simultaneous denial of all the propositions which are values of fx. In this way we arrive at the proposition which is ordinarily represented in mathematical logic by the words 'fx is false for all values of x'. The negation of this would be the proposition 'there is at least one x for which fx is true' which is represented by '$(\exists x).fx$'. If we had started with not-fx instead of fx we should have arrived at the proposition 'fx is true for all values of x' which is represented by '$(x).fx$'. Wittgenstein's method of dealing with general propositions [i.e. '$(x).fx$' and '$(\exists x).fx$'] differs from previous methods by the fact that the generality comes only in specifying the set of propositions concerned, and when this has been done the building up of truth-functions proceeds exactly as it would in the case of a finite number of enumerated arguments p, q, r, ...

Mr Wittgenstein's explanation of his symbolism at this point is not quite fully given in the text. The symbol he uses is

$$[\bar{p}, \bar{\xi}, N(\bar{\xi})].$$

The following is the explanation of this symbol:

$\bar{p}$ stands for all atomic propositions.
$\bar{\xi}$ stands for any set of propositions.
$N(\bar{\xi})$ stands for the negation of all the propositions making up $\bar{\xi}$.

The whole symbol $[\bar{p}, \bar{\xi}, N(\bar{\xi})]$ means whatever can be obtained by taking any selection of atomic propositions, negating them all, then taking any selection of the set of propositions now obtained, together with any of the originals—and so on indefinitely. This is, he says, the general truth-function and also the general form of proposition. What is meant is somewhat less complicated than it sounds. The symbol is intended to describe a process by the help of which, given the atomic propositions, all others can be manufactured. The process depends upon:

(*a*) Sheffer's proof that all truth-functions can be obtained out of simultaneous negation, i.e. out of 'not-p and not-q';

(*b*) Mr Wittgenstein's theory of the derivation of general propositions from conjunctions and disjunctions;

(*c*) The assertion that a proposition can only occur in another proposition as argument to a truth-function.
Given these three foundations, it follows that all propositions which are not atomic can be derived from such as are, by a uniform process, and it is this process which is indicated by Mr Wittgenstein's symbol.

From this uniform method of construction we arrive at an amazing simplification of the theory of inference, as well as a definition of the sort of propositions that belong to logic. The method of generation which has just been described enables Wittgenstein to say that all propositions can be constructed in the above manner from atomic propositions, and in this way the totality of propositions is defined. (The apparent exceptions which we mentioned above are dealt with in a manner which we shall consider later.) Wittgenstein is enabled to assert that propositions are all that follows from the totality of atomic propositions (together with the fact that it is the totality of them); that a proposition is always a truth-function of atomic propositions; and that if *p* follows from *q* the meaning of *p* is contained in the meaning of *q*, from which of course it results that nothing can be deduced from an atomic proposition. All the propositions of logic, he maintains, are tautologies, such, for example, as '*p* or not-*p*'.

The fact that nothing can be deduced from an atomic proposition has interesting applications, for example, to causality. There cannot, in Wittgenstein's logic, be any such thing as a causal nexus. 'The events of the future', he says, '*cannot* be inferred from those of the present. Superstition is the belief in the causal nexus.' That the sun will rise to-morrow is a hypothesis. We do not in fact know whether it will rise, since there is no compulsion according to which one thing must happen because another happens.

Let us now take up another subject—that of names. In Wittgenstein's theoretical logical language, names are only given to simples. We do not give two names to one thing, or one name to two things. There is no way whatever, according to him, by which we can describe the totality of things that can be named, in other words, the totality of what there is in the world. In order to be able to do this we should have to know of some property which must belong to every thing by a logical necessity. It has been

sought to find such a property in self-identity, but the conception of identity is subjected by Wittgenstein to a destructive criticism from which there seems no escape. The definition of identity by means of the identity of indiscernibles is rejected, because the identity of indiscernibles appears to be not a logically necessary principle. According to this principle x is identical with y if every property of x is a property of y, but it would, after all, be logically possible for two things to have exactly the same properties. If this does not in fact happen that is an accidental characteristic of the world, not a logically necessary characteristic, and accidental characteristics of the world must, of course, not be admitted into the structure of logic. Mr Wittgenstein accordingly banishes identity and adopts the convention that different letters are to mean different things. In practice, identity is needed as between a name and a description or between two descriptions. It is needed for such propositions as 'Socrates is the philosopher who drank the hemlock', or 'The even prime is the next number after 1'. For such uses of identity it is easy to provide on Wittgenstein's system.

The rejection of identity removes one method of speaking of the totality of things, and it will be found that any other method that may be suggested is equally fallacious: so, at least, Wittgenstein contends and, I think, rightly. This amounts to saying that 'object' is a pseudo-concept. To say 'x is an object' is to say nothing. It follows from this that we cannot make such statements as 'there are more than three objects in the world', or 'there are an infinite number of objects in the world'. Objects can only be mentioned in connexion with some definite property. We can say 'there are more than three objects which are human', or 'there are more than three objects which are red', for in these statements the word 'object' can be replaced by a variable in the language of logic, the variable being one which satisfies in the first case the function 'x is human'; in the second the function 'x is red'. But when we attempt to say 'there are more than three objects', this substitution of the variable for the word 'object' becomes impossible, and the proposition is therefore seen to be meaningless.

We here touch one instance of Wittgenstein's fundamental thesis, that it is impossible to say anything about the world as a whole, and that whatever can be said has to be about bounded portions of the world. This view may have been originally suggested by notation, and if so, that is much in its favour, for a good

notation has a subtlety and suggestiveness which at times make it seem almost like a live teacher. Notational irregularities are often the first sign of philosophical errors, and a perfect notation would be a substitute for thought. But although notation may have first suggested to Mr Wittgenstein the limitation of logic to things within the world as opposed to the world as a whole, yet the view, once suggested, is seen to have much else to recommend it. Whether it is ultimately true I do not, for my part, profess to know. In this Introduction I am concerned to expound it, not to pronounce upon it. According to this view we could only say things about the world as a whole if we could get outside the world, if, that is to say, it ceased to be for us the whole world. Our world may be bounded for some superior being who can survey it from above, but for us, however finite it may be, it cannot have a boundary, since it has nothing outside it. Wittgenstein uses, as an analogy, the field of vision. Our field of vision does not, for us, have a visual boundary, just because there is nothing outside it, and in like manner our logical world has no logical boundary because our logic knows of nothing outside it. These considerations lead him to a somewhat curious discussion of Solipsism. Logic, he says, fills the world. The boundaries of the world are also its boundaries. In logic, therefore, we cannot say, there is this and this in the world, but not that, for to say so would apparently presuppose that we exclude certain possibilities, and this cannot be the case, since it would require that logic should go beyond the boundaries of the world as if it could contemplate these boundaries from the other side also. What we cannot think we cannot think, therefore we also cannot say what we cannot think.

This, he says, gives the key to Solipsism. What Solipsism intends is quite correct, but this cannot be said, it can only be shown. That the world is *my* world appears in the fact that the boundaries of language (the only language I understand) indicate the boundaries of my world. The metaphysical subject does not belong to the world but is a boundary of the world.

We must take up next the question of molecular propositions which are at first sight not truth-functions of the propositions that they contain, such, for example, as '*A* believes *p*'.

Wittgenstein introduces this subject in the statement of his position, namely, that all molecular functions are truth-functions. He says (5.54): 'In the general propositional form, propositions

occur in a proposition only as bases of truth-operations.' At first sight, he goes on to explain, it seems as if a proposition could also occur in other ways, e.g. '*A* believes *p*'. Here it seems superficially as if the proposition *p* stood in a sort of relation to the object *A*. 'But it is clear that "*A* believes that *p*", "*A* thinks *p*", "*A* says *p*" are of the form "'*p*' says *p*"; and here we have no co-ordination of a fact and an object, but a co-ordination of facts by means of a co-ordination of their objects' (5.542).

What Mr Wittgenstein says here is said so shortly that its point is not likely to be clear to those who have not in mind the controversies with which he is concerned. The theory with which he is disagreeing will be found in my articles on the nature of truth and falsehood in *Philosophical Essays* and *Proceedings of the Aristotelian Society*, 1906–7. The problem at issue is the problem of the logical form of belief, i.e. what is the schema representing what occurs when a man believes. Of course, the problem applies not only to belief, but also to a host of other mental phenomena which may be called propositional attitudes: doubting, considering, desiring, etc. In all these cases it seems natural to express the phenomenon in the form '*A* doubts *p*', '*A* desires *p*', etc., which makes it appear as though we were dealing with a relation between a person and a proposition. This cannot, of course, be the ultimate analysis, since persons are fictions and so are propositions, except in the sense in which they are facts on their own account. A proposition, considered as a fact on its own account, may be a set of words which a man says over to himself, or a complex image, or train of images passing through his mind, or a set of incipient bodily movements. It may be any one of innumerable different things. The proposition as a fact on its own account, for example the actual set of words the man pronounces to himself, is not relevant to logic. What is relevant to logic is that common element among all these facts, which enables him, as we say, to *mean* the fact which the proposition asserts. To psychology, of course, more is relevant; for a symbol does not mean what it symbolizes in virtue of a logical relation alone, but in virtue also of a psychological relation of intention, or association, or what-not. The psychological part of meaning, however, does not concern the logician. What does concern him in this problem of belief is the logical schema. It is clear that, when a person believes a proposition, the person, considered as a metaphysical subject, does not have to be assumed in order to

explain what is happening. What has to be explained is the relation between the set of words which is the proposition considered as a fact on its own account, and the 'objective' fact which makes the proposition true or false. This reduces ultimately to the question of the meaning of propositions, that is to say, the meaning of propositions is the only non-psychological portion of the problem involved in the analysis of belief. This problem is simply one of a relation of two facts, namely, the relation between the series of words used by the believer and the fact which makes these words true or false. The series of words is a fact just as much as what makes it true or false is a fact. The relation between these two facts is not unanalysable, since the meaning of a proposition results from the meaning of its constituent words. The meaning of the series of words which is a proposition is a function of the meanings of the separate words. Accordingly, the proposition as a whole does not really enter into what has to be explained in explaining the meaning of a proposition. It would perhaps help to suggest the point of view which I am trying to indicate, to say that in the cases we have been considering the proposition occurs as a fact, not as a proposition. Such a statement, however, must not be taken too literally. The real point is that in believing, desiring, etc., what is logically fundamental is the relation of a proposition, *considered as a fact*, to the fact which makes it true or false, and that this relation of two facts is reducible to a relation of their constituents. Thus the proposition does not occur at all in the same sense in which it occurs in a truth-function.

There are some respects, in which, as it seems to me, Mr Wittgenstein's theory stands in need of greater technical development. This applies in particular to his theory of number (6.02 ff.) which, as it stands, is only capable of dealing with finite numbers. No logic can be considered adequate until it has been shown to be capable of dealing with transfinite numbers. I do not think there is anything in Mr Wittgenstein's system to make it impossible for him to fill this lacuna.

More interesting than such questions of comparative detail is Mr Wittgenstein's attitude towards the mystical. His attitude upon this grows naturally out of his doctrine in pure logic, according to which the logical proposition is a picture (true or false) of the fact, and has in common with the fact a certain structure. It is this common structure which makes it capable of being a picture of

the fact, but the structure cannot itself be put into words, since it is a structure *of* words, as well as of the facts to which they refer. Everything, therefore, which is involved in the very idea of the expressiveness of language must remain incapable of being expressed in language, and is, therefore, inexpressible in a perfectly precise sense. This inexpressible contains, according to Mr Wittgenstein, the whole of logic and philosophy. The right method of teaching philosophy, he says, would be to confine oneself to propositions of the sciences, stated with all possible clearness and exactness, leaving philosophical assertions to the learner, and proving to him, whenever he made them, that they are meaningless. It is true that the fate of Socrates might befall a man who attempted this method of teaching, but we are not to be deterred by that fear, if it is the only right method. It is not this that causes some hesitation in accepting Mr Wittgenstein's position, in spite of the very powerful arguments which he brings to its support. What causes hesitation is the fact that, after all, Mr Wittgenstein manages to say a good deal about what cannot be said, thus suggesting to the sceptical reader that possibly there may be some loophole through a hierarchy of languages, or by some other exit. The whole subject of ethics, for example, is placed by Mr Wittgenstein in the mystical, inexpressible region. Nevertheless he is capable of conveying his ethical opinions. His defence would be that what he calls the mystical can be shown, although it cannot be said. It may be that this defence is adequate, but, for my part, I confess that it leaves me with a certain sense of intellectual discomfort.

There is one purely logical problem in regard to which these difficulties are peculiarly acute. I mean the problem of generality. In the theory of generality it is necessary to consider all propositions of the form fx where fx is a given propositional function. This belongs to the part of logic which can be expressed, according to Mr Wittgenstein's system. But the totality of possible values of x which might seem to be involved in the totality of propositions of the form fx is not admitted by Mr Wittgenstein among the things that can be spoken of, for this is no other than the totality of things in the world, and thus involves the attempt to conceive the world as a whole; 'the feeling of the world as a bounded whole is the mystical'; hence the totality of the values of x is mystical (6.45). This is expressly argued when Mr

Wittgenstein denies that we can make propositions as to how many things there are in the world, as for example, that there are more than three.

These difficulties suggest to my mind some such possibility as this: that every language has, as Mr Wittgenstein says, a structure concerning which, *in the language*, nothing can be said, but that there may be another language dealing with the structure of the first language, and having itself a new structure, and that to this hierarchy of languages there may be no limit. Mr Wittgenstein would of course reply that his whole theory is applicable unchanged to the totality of such languages. The only retort would be to deny that there is any such totality. The totalities concerning which Mr Wittgenstein holds that it is impossible to speak logically are nevertheless thought by him to exist, and are the subject-matter of his mysticism. The totality resulting from our hierarchy would be not merely logically inexpressible, but a fiction, a mere delusion, and in this way the supposed sphere of the mystical would be abolished. Such an hypothesis is very difficult, and I can see objections to it which at the moment I do not know how to answer. Yet I do not see how any easier hypothesis can escape from Mr Wittgenstein's conclusions. Even if this very difficult hypothesis should prove tenable, it would leave untouched a very large part of Mr Wittgenstein's theory, though possibly not the part upon which he himself would wish to lay most stress. As one with a long experience of the difficulties of logic and of the deceptiveness of theories which seem irrefutable, I find myself unable to be sure of the rightness of a theory, merely on the ground that I cannot see any point on which it is wrong. But to have constructed a theory of logic which is not at any point obviously wrong is to have achieved a work of extraordinary difficulty and importance. This merit, in my opinion, belongs to Mr Wittgenstein's book, and makes it one which no serious philosopher can afford to neglect.

BERTRAND RUSSELL

May 1922

LOGISCH-PHILOSOPHISCHE ABHANDLUNG

TRACTATUS LOGICO-PHILOSOPHICUS

Dem Andenken meines Freundes
David H. Pinsent
gewidmet

Dedicated to the memory of my friend
David H. Pinsent

Motto: ... und alles, was man weiß, nicht bloß rauschen und brausen gehört hat, läßt sich in drei Worten sagen.
Kürnberger

Motto: ... and whatever a man knows, whatever is not men rumbling and roaring that he has heard, can be said in three words.
Kürnberger

ERRATA

p.1 l.1 of *Motto:* for '*men*' read '*mere*'

p.35 l.2 of 3·3442: delete 'not'

LOGISCH-PHILOSOPHISCHE ABHANDLUNG

VORWORT

Dieses Buch wird vielleicht nur der verstehen, der die Gedanken, die darin ausgedrückt sind — oder doch ähnliche Gedanken — schon selbst einmal gedacht hat.— Es ist also kein Lehrbuch.— Sein Zweck wäre erreicht, wenn es Einem, der es mit Verständnis liest, Vergnügen bereitete.

Das Buch behandelt die philosophischen Probleme und zeigt — wie ich glaube —, daß die Fragestellung dieser Probleme auf dem Mißverständnis der Logik unserer Sprache beruht. Man könnte den ganzen Sinn des Buches etwa in die Worte fassen: Was sich überhaupt sagen läßt, läßt sich klar sagen; und wovon man nicht reden kann, darüber muß man schweigen.

Das Buch will also dem Denken eine Grenze ziehen, oder vielmehr — nicht dem Denken, sondern dem Ausdruck der Gedanken: Denn um dem Denken eine Grenze zu ziehen, müßten wir beide Seiten dieser Grenze denken können (wir müßten also denken können, was sich nicht denken läßt).

Die Grenze wird also nur in der Sprache gezogen werden können und was jenseits der Grenze liegt, wird einfach Unsinn sein.

Wieweit meine Bestrebungen mit denen anderer Philosophen zusammenfallen, will ich nicht beurteilen. Ja, was ich hier geschrieben habe, macht im Einzelnen überhaupt nicht den Anspruch auf Neuheit; und darum gebe ich auch keine Quellen an, weil es mir gleichgültig ist, ob das, was ich gedacht habe, vor mir schon ein anderer gedacht hat.

Nur das will ich erwähnen, daß ich den großartigen Werken Freges und den Arbeiten meines Freundes Herrn Bertrand Russell einen großen Teil der Anregung zu meinen Gedanken schulde.

Wenn diese Arbeit einen Wert hat, so besteht er in zweierlei. Erstens darin, daß in ihr Gedanken ausgedrückt sind, und dieser

TRACTATUS LOGICO-PHILOSOPHICUS

PREFACE

Perhaps this book will be understood only by someone who has himself already had the thoughts that are expressed in it—or at least similar thoughts.—So it is not a textbook.—Its purpose would be achieved if it gave pleasure to one person who read and understood it.

The book deals with the problems of philosophy, and shows, I believe, that the reason why these problems are posed is that the logic of our language is misunderstood. The whole sense of the book might be summed up in the following words: what can be said at all can be said clearly, and what we cannot talk about we must pass over in silence.

Thus the aim of the book is to draw a limit to thought, or rather—not to thought, but to the expression of thoughts: for in order to be able to draw a limit to thought, we should have to find both sides of the limit thinkable (i.e. we should have to be able to think what cannot be thought).

It will therefore only be in language that the limit can be drawn, and what lies on the other side of the limit will simply be nonsense.

I do not wish to judge how far my efforts coincide with those of other philosophers. Indeed, what I have written here makes no claim to novelty in detail, and the reason why I give no sources is that it is a matter of indifference to me whether the thoughts that I have had have been anticipated by someone else.

I will only mention that I am indebted to Frege's great works and to the writings of my friend Mr Bertrand Russell for much of the stimulation of my thoughts.

If this work has any value, it consists in two things: the first is that thoughts are expressed in it, and on this score the better the thoughts are expressed—the more the nail has been hit on

Wert wird umso größer sein, je besser die Gedanken ausgedrückt sind. Je mehr der Nagel auf den Kopf getroffen ist.— Hier bin ich mir bewußt, weit hinter dem Möglichen zurückgeblieben zu sein. Einfach darum, weil meine Kraft zur Bewältigung der Aufgabe zu gering ist.— Mögen andere kommen und es besser machen.

Dagegen scheint mir die Wahrheit der hier mitgeteilten Gedanken unantastbar und definitiv. Ich bin also der Meinung, die Probleme im Wesentlichen endgültig gelöst zu haben. Und wenn ich mich hierin nicht irre, so besteht nun der Wert dieser Arbeit zweitens darin, daß sie zeigt, wie wenig damit getan ist, daß diese Probleme gelöst sind.

L. W.

Wien, 1918

the head—the greater will be its value.—Here I am conscious of having fallen a long way short of what is possible. Simply because my powers are too slight for the accomplishment of the task.—May others come and do it better.

On the other hand the *truth* of the thoughts that are here communicated seems to me unassailable and definitive. I therefore believe myself to have found, on all essential points, the final solution of the problems. And if I am not mistaken in this belief, then the second thing in which the value of this work consists is that it shows how little is achieved when these problems are solved.

L. W.

Vienna, 1918

1* Die Welt ist alles, was der Fall ist.

1.1 Die Welt ist die Gesamtheit der Tatsachen, nicht der Dinge.

1.11 Die Welt ist durch die Tatsachen bestimmt und dadurch, daß es a l l e Tatsachen sind.

1.12 Denn, die Gesamtheit der Tatsachen bestimmt, was der Fall ist und auch, was alles nicht der Fall ist.

1.13 Die Tatsachen im logischen Raum sind die Welt.

1.2 Die Welt zerfällt in Tatsachen.

1.21 Eines kann der Fall sein oder nicht der Fall sein und alles übrige gleich bleiben.

2 Was der Fall ist, die Tatsache, ist das Bestehen von Sachverhalten.

2.01 Der Sachverhalt ist eine Verbindung von Gegenständen (Sachen, Dingen).

2.011 Es ist dem Ding wesentlich, der Bestandteil eines Sachverhaltes sein zu können.

2.012 In der Logik ist nichts zufällig: Wenn das Ding im Sachverhalt vorkommen k a n n, so muß die Möglichkeit des Sachverhaltes im Ding bereits präjudiziert sein.

2.0121 Es erschiene gleichsam als Zufall, wenn dem Ding, das allein für sich bestehen könnte, nachträglich eine Sachlage passen würde.

* Die Decimalzahlen als Nummern der einzelnen Sätze deuten das logische Gewicht der Sätze an, den Nachdruck, der auf ihnen in meiner Darstellung liegt. Die Sätze n.1, n.2, n.3, etc. sind Bemerkungen zum Satze No. n; die Sätze n.m1, n.m2, etc. Bemerkungen zum Satze No. n.m; und so weiter.

1* The world is all that is the case.

1.1 The world is the totality of facts, not of things.

1.11 The world is determined by the facts, and by their being *all* the facts.

1.12 For the totality of facts determines what is the case, and also whatever is not the case.

1.13 The facts in logical space are the world.

1.2 The world divides into facts.

1.21 Each item can be the case or not the case while everything else remains the same.

2 What is the case—a fact—is the existence of states of affairs.

2.01 A state of affairs (a state of things) is a combination of objects (things).

2.011 It is essential to things that they should be possible constituents of states of affairs.

2.012 In logic nothing is accidental: if a thing *can* occur in a state of affairs, the possibility of the state of affairs must be written into the thing itself.

2.0121 It would seem to be a sort of accident, if it turned out that a situation would fit a thing that could already exist entirely on its own.

* The decimal numbers assigned to the individual propositions indicate the logical importance of the propositions, the stress laid on them in my exposition. The propositions *n*.1, *n*.2, *n*.3, etc. are comments on proposition no. *n*; the propositions *n*.*m*1, *n*.*m*2, etc. are comments on proposition no. *n*.*m*; and so on.

Wenn die Dinge in Sachverhalten vorkommen können, so muß dies schon in ihnen liegen.

(Etwas Logisches kann nicht nur-möglich sein. Die Logik handelt von jeder Möglichkeit und alle Möglichkeiten sind ihre Tatsachen.)

Wie wir uns räumliche Gegenstände überhaupt nicht außerhalb des Raumes, zeitliche nicht außerhalb der Zeit denken können, so können wir uns *keinen* Gegenstand außerhalb der Möglichkeit seiner Verbindung mit anderen denken.

Wenn ich mir den Gegenstand im Verbande des Sachverhalts denken kann, so kann ich ihn nicht außerhalb der *Möglichkeit* dieses Verbandes denken.

2.0122 Das Ding ist selbständig, insofern es in allen *möglichen* Sachlagen vorkommen kann, aber diese Form der Selbständigkeit ist eine Form des Zusammenhangs mit dem Sachverhalt, eine Form der Unselbständigkeit. (Es ist unmöglich, daß Worte in zwei verschiedenen Weisen auftreten, allein und im Satz.)

2.0123 Wenn ich den Gegenstand kenne, so kenne ich auch sämtliche Möglichkeiten seines Vorkommens in Sachverhalten.

(Jede solche Möglichkeit muß in der Natur des Gegenstandes liegen.)

Es kann nicht nachträglich eine neue Möglichkeit gefunden werden.

2.01231 Um einen Gegenstand zu kennen, muß ich zwar nicht seine externen — aber ich muß alle seine internen Eigenschaften kennen.

2.0124 Sind alle Gegenstände gegeben, so sind damit auch alle *möglichen* Sachverhalte gegeben.

2.013 Jedes Ding ist, gleichsam, in einem Raume möglicher Sachverhalte. Diesen Raum kann ich mir leer denken, nicht aber das Ding ohne den Raum.

2.0131 Der räumliche Gegenstand muß im unendlichen Raume liegen. (Der Raumpunkt ist eine Argumentstelle.)

Der Fleck im Gesichtsfeld muß zwar nicht rot sein, aber eine Farbe muß er haben: er hat sozusagen den Far-

If things can occur in states of affairs, this possibility must be in them from the beginning.

(Nothing in the province of logic can be merely possible. Logic deals with every possibility and all possibilities are its facts.)

Just as we are quite unable to imagine spatial objects outside space or temporal objects outside time, so too there is *no* object that we can imagine excluded from the possibility of combining with others.

If I can imagine objects combined in states of affairs, I cannot imagine them excluded from the *possibility* of such combinations.

2.0122 Things are independent in so far as they can occur in all *possible* situations, but this form of independence is a form of connexion with states of affairs, a form of dependence. (It is impossible for words to appear in two different rôles: by themselves, and in propositions.)

2.0123 If I know an object I also know all its possible occurrences in states of affairs.

(Every one of these possibilities must be part of the nature of the object.)

A new possibility cannot be discovered later.

2.01231 If I am to know an object, though I need not know its external properties, I must know all its internal properties.

2.0124 If all objects are given, then at the same time all *possible* states of affairs are also given.

2.013 Each thing is, as it were, in a space of possible states of affairs. This space I can imagine empty, but I cannot imagine the thing without the space.

2.0131 A spatial object must be situated in infinite space. (A spatial point is an argument-place.)

A speck in the visual field, though it need not be red, must have some colour: it is, so to speak, surrounded by

benraum um sich. Der Ton muß *eine* Höhe haben, der Gegenstand des Tastsinnes *eine* Härte, usw.

2.014 Die Gegenstände enthalten die Möglichkeit aller Sachlagen.

2.0141 Die Möglichkeit seines Vorkommens in Sachverhalten ist die Form des Gegenstandes.

2.02 Der Gegenstand ist einfach.

2.0201 Jede Aussage über Komplexe läßt sich in eine Aussage über deren Bestandteile und in diejenigen Sätze zerlegen, welche die Komplexe vollständig beschreiben.

2.021 Die Gegenstände bilden die Substanz der Welt. Darum können sie nicht zusammengesetzt sein.

2.0211 Hätte die Welt keine Substanz, so würde, ob ein Satz Sinn hat, davon abhängen, ob ein anderer Satz wahr ist.

2.0212 Es wäre dann unmöglich, ein Bild der Welt (wahr oder falsch) zu entwerfen.

2.022 Es ist offenbar, daß auch eine von der wirklichen noch so verschieden gedachte Welt Etwas — eine Form — mit der wirklichen gemein haben muß.

2.023 Diese feste Form besteht eben aus den Gegenständen.

2.0231 Die Substanz der Welt *kann* nur eine Form und keine materiellen Eigenschaften bestimmen. Denn diese werden erst durch die Sätze dargestellt — erst durch die Konfiguration der Gegenstände gebildet.

2.0232 Beiläufig gesprochen: Die Gegenstände sind farblos.

2.0233 Zwei Gegenstände von der gleichen logischen Form sind — abgesehen von ihren externen Eigenschaften — von einander nur dadurch unterschieden, daß sie verschieden sind.

2.02331 Entweder ein Ding hat Eigenschaften, die kein anderes hat, dann kann man es ohneweiteres durch eine Beschreibung aus den anderen herausheben, und darauf hinweisen; oder aber es gibt mehrere Dinge, die ihre sämtlichen Eigenschaften gemeinsam haben, dann ist es überhaupt unmöglich auf eines von ihnen zu zeigen.

colour-space. Notes must have *some* pitch, objects of the sense of touch *some* degree of hardness, and so on.

2.014 Objects contain the possibility of all situations.

2.0141 The possibility of its occurring in states of affairs is the form of an object.

2.02 Objects are simple.

2.0201 Every statement about complexes can be resolved into a statement about their constituents and into the propositions that describe the complexes completely.

2.021 Objects make up the substance of the world. That is why they cannot be composite.

2.0211 If the world had no substance, then whether a proposition had sense would depend on whether another proposition was true.

2.0212 In that case we could not sketch any picture of the world (true or false).

2.022 It is obvious that an imagined world, however different it may be from the real one, must have *something*—a form—in common with it.

2.023 Objects are just what constitute this unalterable form.

2.0231 The substance of the world *can* only determine a form, and not any material properties. For it is only by means of propositions that material properties are represented—only by the configuration of objects that they are produced.

2.0232 In a manner of speaking, objects are colourless.

2.0233 If two objects have the same logical form, the only distinction between them, apart from their external properties, is that they are different.

2.02331 Either a thing has properties that nothing else has, in which case we can immediately use a description to distinguish it from the others and refer to it; or, on the other hand, there are several things that have the whole set of their properties in common, in which case it is quite impossible to indicate one of them.

Denn, ist das Ding durch nichts hervorgehoben, so kann ich es nicht hervorheben, denn sonst ist es eben hervorgehoben.

2.024 Die Substanz ist das, was unabhängig von dem, was der Fall ist, besteht.

2.025 Sie ist Form und Inhalt.

2.0251 Raum, Zeit und Farbe (Färbigkeit) sind Formen der Gegenstände.

2.026 Nur wenn es Gegenstände gibt, kann es eine feste Form der Welt geben.

2.027 Das Feste, das Bestehende und der Gegenstand sind Eins.

2.0271 Der Gegenstand ist das Feste, Bestehende; die Konfiguration ist das Wechselnde, Unbeständige.

2.0272 Die Konfiguration der Gegenstände bildet den Sachverhalt.

2.03 Im Sachverhalt hängen die Gegenstände ineinander, wie die Glieder einer Kette.

2.031 Im Sachverhalt verhalten sich die Gegenstände in bestimmter Art und Weise zueinander.

2.032 Die Art und Weise, wie die Gegenstände im Sachverhalt zusammenhängen, ist die Struktur des Sachverhaltes.

2.033 Die Form ist die Möglichkeit der Struktur.

2.034 Die Struktur der Tatsache besteht aus den Strukturen der Sachverhalte.

2.04 Die Gesamtheit der bestehenden Sachverhalte ist die Welt.

2.05 Die Gesamtheit der bestehenden Sachverhalte bestimmt auch, welche Sachverhalte nicht bestehen.

2.06 Das Bestehen und Nichtbestehen von Sachverhalten ist die Wirklichkeit.

(Das Bestehen von Sachverhalten nennen wir auch eine positive, das Nichtbestehen eine negative Tatsache.)

2.061 Die Sachverhalte sind von einander unabhängig.

2.062 Aus dem Bestehen oder Nichtbestehen eines Sachverhaltes kann nicht auf das Bestehen oder Nichtbestehen eines anderen geschlossen werden.

For if there is nothing to distinguish a thing, I cannot distinguish it, since otherwise it would be distinguished after all.

2.024 Substance is what subsists independently of what is the case.

2.025 It is form and content.

2.0251 Space, time, and colour (being coloured) are forms of objects.

2.026 There must be objects, if the world is to have an unalterable form.

2.027 Objects, the unalterable, and the subsistent are one and the same.

2.0271 Objects are what is unalterable and subsistent; their configuration is what is changing and unstable.

2.0272 The configuration of objects produces states of affairs.

2.03 In a state of affairs objects fit into one another like the links of a chain.

2.031 In a state of affairs objects stand in a determinate relation to one another.

2.032 The determinate way in which objects are connected in a state of affairs is the structure of the state of affairs.

2.033 Form is the possibility of structure.

2.034 The structure of a fact consists of the structures of states of affairs.

2.04 The totality of existing states of affairs is the world.

2.05 The totality of existing states of affairs also determines which states of affairs do not exist.

2.06 The existence and non-existence of states of affairs is reality.

(We also call the existence of states of affairs a positive fact, and their non-existence a negative fact.)

2.061 States of affairs are independent of one another.

2.062 From the existence or non-existence of one state of affairs it is impossible to infer the existence or non-existence of another.

2.063 Die gesamte Wirklichkeit ist die Welt.

2.1 Wir machen uns Bilder der Tatsachen.

2.11 Das Bild stellt die Sachlage im logischen Raume, das Bestehen und Nichtbestehen von Sachverhalten, vor.

2.12 Das Bild ist ein Modell der Wirklichkeit.

2.13 Den Gegenständen entsprechen im Bilde die Elemente des Bildes.

2.131 Die Elemente des Bildes vertreten im Bild die Gegenstände.

2.14 Das Bild besteht darin, daß sich seine Elemente in bestimmter Art und Weise zu einander verhalten.

2.141 Das Bild ist eine Tatsache.

2.15 Daß sich die Elemente des Bildes in bestimmter Art und Weise zu einander verhalten, stellt vor, daß sich die Sachen so zu einander verhalten.

Dieser Zusammenhang der Elemente des Bildes heiße seine Struktur und ihre Möglichkeit seine Form der Abbildung.

2.151 Die Form der Abbildung ist die Möglichkeit, daß sich die Dinge so zu einander verhalten, wie die Elemente des Bildes.

2.1511 Das Bild ist *so* mit der Wirklichkeit verknüpft; es reicht bis zu ihr.

2.1512 Es ist wie ein Maßstab an die Wirklichkeit angelegt.

2.15121 Nur die äußersten Punkte der Teilstriche *berühren* den zu messenden Gegenstand.

2.1513 Nach dieser Auffassung gehört also zum Bilde auch noch die abbildende Beziehung, die es zum Bild macht.

2.1514 Die abbildende Beziehung besteht aus den Zuordnungen der Elemente des Bildes und der Sachen.

2.1515 Diese Zuordnungen sind gleichsam die Fühler der Bildelemente, mit denen das Bild die Wirklichkeit berührt.

2.16 Die Tatsache muß, um Bild zu sein, etwas mit dem Abgebildeten gemeinsam haben.

2.161 In Bild und Abgebildetem muß etwas identisch sein, damit das eine überhaupt ein Bild des anderen sein kann.

2.063 The sum-total of reality is the world.

2.1 We picture facts to ourselves.

2.11 A picture presents a situation in logical space, the existence and non-existence of states of affairs.

2.12 A picture is a model of reality.

2.13 In a picture objects have the elements of the picture corresponding to them.

2.131 In a picture the elements of the picture are the representatives of objects.

2.14 What constitutes a picture is that its elements are related to one another in a determinate way.

2.141 A picture is a fact.

2.15 The fact that the elements of a picture are related to one another in a determinate way represents that things are related to one another in the same way.

Let us call this connexion of its elements the structure of the picture, and let us call the possibility of this structure the pictorial form of the picture.

2.151 Pictorial form is the possibility that things are related to one another in the same way as the elements of the picture.

2.1511 *That* is how a picture is attached to reality; it reaches right out to it.

2.1512 It is laid against reality like a measure.

2.15121 Only the end-points of the graduating lines actually *touch* the object that is to be measured.

2.1513 So a picture, conceived in this way, also includes the pictorial relationship, which makes it into a picture.

2.1514 The pictorial relationship consists of the correlations of the picture's elements with things.

2.1515 These correlations are, as it were, the feelers of the picture's elements, with which the picture touches reality.

2.16 If a fact is to be a picture, it must have something in common with what it depicts.

2.161 There must be something identical in a picture and what it depicts, to enable the one to be a picture of the other at all.

2.17 Was das Bild mit der Wirklichkeit gemein haben muß, um sie auf seine Art und Weise — richtig oder falsch — abbilden zu können, ist seine Form der Abbildung.

2.171 Das Bild kann jede Wirklichkeit abbilden, deren Form es hat.

Das räumliche Bild alles Räumliche, das farbige alles Farbige, etc.

2.172 Seine Form der Abbildung aber kann das Bild nicht abbilden; es weist sie auf.

2.173 Das Bild stellt sein Objekt von außerhalb dar (sein Standpunkt ist seine Form der Darstellung), darum stellt das Bild sein Objekt richtig oder falsch dar.

2.174 Das Bild kann sich aber nicht außerhalb seiner Form der Darstellung stellen.

2.18 Was jedes Bild, welcher Form immer, mit der Wirklichkeit gemein haben muß, um sie überhaupt — richtig oder falsch — abbilden zu können, ist die logische Form, das ist, die Form der Wirklichkeit.

2.181 Ist die Form der Abbildung die logische Form, so heißt das Bild das logische Bild.

2.182 Jedes Bild ist a u c h ein logisches. (Dagegen ist z. B. nicht jedes Bild ein räumliches.)

2.19 Das logische Bild kann die Welt abbilden.

2.2 Das Bild hat mit dem Abgebildeten die logische Form der Abbildung gemein.

2.201 Das Bild bildet die Wirklichkeit ab, indem es eine Möglichkeit des Bestehens und Nichtbestehens von Sachverhalten darstellt.

2.202 Das Bild stellt eine mögliche Sachlage im logischen Raume dar.

2.203 Das Bild enthält die Möglichkeit der Sachlage, die es darstellt.

2.21 Das Bild stimmt mit der Wirklichkeit überein oder nicht; es ist richtig oder unrichtig, wahr oder falsch.

2.22 Das Bild stellt dar, was es darstellt, unabhängig von seiner Wahr- oder Falschheit, durch die Form der Abbildung.

2.17 What a picture must have in common with reality, in order to be able to depict it—correctly or incorrectly—in the way it does, is its pictorial form.

2.171 A picture can depict any reality whose form it has.

A spatial picture can depict anything spatial, a coloured one anything coloured, etc.

2.172 A picture cannot, however, depict its pictorial form: it displays it.

2.173 A picture represents its subject from a position outside it. (Its standpoint is its representational form.) That is why a picture represents its subject correctly or incorrectly.

2.174 A picture cannot, however, place itself outside its representational form.

2.18 What any picture, of whatever form, must have in common with reality, in order to be able to depict it—correctly or incorrectly—in any way at all, is logical form, i.e. the form of reality.

2.181 A picture whose pictorial form is logical form is called a logical picture.

2.182 Every picture is *at the same time* a logical one. (On the other hand, not every picture is, for example, a spatial one.)

2.19 Logical pictures can depict the world.

2.2 A picture has logico-pictorial form in common with what it depicts.

2.201 A picture depicts reality by representing a possibility of existence and non-existence of states of affairs.

2.202 A picture represents a possible situation in logical space.

2.203 A picture contains the possibility of the situation that it represents.

2.21 A picture agrees with reality or fails to agree; it is correct or incorrect, true or false.

2.22 What a picture represents it represents independently of its truth or falsity, by means of its pictorial form.

2.221 Was das Bild darstellt, ist sein Sinn.

2.222 In der Übereinstimmung oder Nichtübereinstimmung seines Sinnes mit der Wirklichkeit besteht seine Wahrheit oder Falschheit.

2.223 Um zu erkennen, ob das Bild wahr oder falsch ist, müssen wir es mit der Wirklichkeit vergleichen.

2.224 Aus dem Bild allein ist nicht zu erkennen, ob es wahr oder falsch ist.

2.225 Ein a priori wahres Bild gibt es nicht.

3 Das logische Bild der Tatsachen ist der Gedanke.

3.001 „Ein Sachverhalt ist denkbar", heißt: Wir können uns ein Bild von ihm machen.

3.01 Die Gesamtheit der wahren Gedanken sind ein Bild der Welt.

3.02 Der Gedanke enthält die Möglichkeit der Sachlage, die er denkt. Was denkbar ist, ist auch möglich.

3.03 Wir können nichts Unlogisches denken, weil wir sonst unlogisch denken müßten.

3.031 Man sagte einmal, daß Gott alles schaffen könne, nur nichts, was den logischen Gesetzen zuwider wäre.— Wir könnten nämlich von einer „unlogischen" Welt nicht *sagen*, wie sie aussähe.

3.032 Etwas „der Logik widersprechendes" in der Sprache darstellen, kann man ebensowenig, wie in der Geometrie eine den Gesetzen des Raumes widersprechende Figur durch ihre Koordinaten darstellen; oder die Koordinaten eines Punktes angeben, welcher nicht existiert.

3.0321 Wohl können wir einen Sachverhalt räumlich darstellen, welcher den Gesetzen der Physik, aber keinen, der den Gesetzen der Geometrie zuwiderliefe.

3.04 Ein a priori richtiger Gedanke wäre ein solcher, dessen Möglichkeit seine Wahrheit bedingte.

3.05 Nur so könnten wir a priori wissen, daß ein Gedanke wahr ist, wenn aus dem Gedanken selbst (ohne Vergleichsobjekt) seine Wahrheit zu erkennen wäre.

3.1 Im Satz drückt sich der Gedanke sinnlich wahrnehmbar aus.

2.221 What a picture represents is its sense.

2.222 The agreement or disagreement of its sense with reality constitutes its truth or falsity.

2.223 In order to tell whether a picture is true or false we must compare it with reality.

2.224 It is impossible to tell from the picture alone whether it is true or false.

2.225 There are no pictures that are true a priori.

3 A logical picture of facts is a thought.

3.001 'A state of affairs is thinkable': what this means is that we can picture it to ourselves.

3.01 The totality of true thoughts is a picture of the world.

3.02 A thought contains the possibility of the situation of which it is the thought. What is thinkable is possible too.

3.03 Thought can never be of anything illogical, since, if it were, we should have to think illogically.

3.031 It used to be said that God could create anything except what would be contrary to the laws of logic.—The truth is that we could not *say* what an 'illogical' world would look like.

3.032 It is as impossible to represent in language anything that 'contradicts logic' as it is in geometry to represent by its co-ordinates a figure that contradicts the laws of space, or to give the co-ordinates of a point that does not exist.

3.0321 Though a state of affairs that would contravene the laws of physics can be represented by us spatially, one that would contravene the laws of geometry cannot.

3.04 If a thought were correct a priori, it would be a thought whose possibility ensured its truth.

3.05 A priori knowledge that a thought was true would be possible only if its truth were recognizable from the thought itself (without anything to compare it with).

3.1 In a proposition a thought finds an expression that can be perceived by the senses.

3.11 Wir benützen das sinnlich wahrnehmbare Zeichen (Laut- oder Schriftzeichen etc.) des Satzes als Projektion der möglichen Sachlage.

Die Projektionsmethode ist das Denken des Satz-Sinnes.

3.12 Das Zeichen, durch welches wir den Gedanken ausdrücken, nenne ich das Satzzeichen. Und der Satz ist das Satzzeichen in seiner projektiven Beziehung zur Welt.

3.13 Zum Satz gehört alles, was zur Projektion gehört; aber nicht das Projizierte.

Also die Möglichkeit des Projizierten, aber nicht dieses selbst.

Im Satz ist also sein Sinn noch nicht enthalten, wohl aber die Möglichkeit ihn auszudrücken.

(„Der Inhalt des Satzes" heißt der Inhalt des sinnvollen Satzes.)

Im Satz ist die Form seines Sinnes enthalten, aber nicht dessen Inhalt.

3.14 Das Satzzeichen besteht darin, daß sich seine Elemente, die Wörter, in ihm auf bestimmte Art und Weise zu einander verhalten.

Das Satzzeichen ist eine Tatsache.

3.141 Der Satz ist kein Wörtergemisch.—(Wie das musikalische Thema kein Gemisch von Tönen.)

Der Satz ist artikuliert.

3.142 Nur Tatsachen können einen Sinn ausdrücken, eine Klasse von Namen kann es nicht.

3.143 Daß das Satzzeichen eine Tatsache ist, wird durch die gewöhnliche Ausdrucksform der Schrift oder des Druckes verschleiert.

Denn im gedruckten Satz z. B. sieht das Satzzeichen nicht wesentlich verschieden aus vom Wort.

(So war es möglich, daß Frege den Satz einen zusammengesetzten Namen nannte.)

3.1431 Sehr klar wird das Wesen des Satzzeichens, wenn wir es uns, statt aus Schriftzeichen, aus räumlichen Gegenständen (etwa Tischen, Stühlen, Büchern) zusammengesetzt denken.

Die gegenseitige räumliche Lage dieser Dinge drückt dann den Sinn des Satzes aus.

3.11 We use the perceptible sign of a proposition (spoken or written, etc.) as a projection of a possible situation.

The method of projection is to think of the sense of the proposition.

3.12 I call the sign with which we express a thought a propositional sign.—And a proposition is a propositional sign in its projective relation to the world.

3.13 A proposition includes all that the projection includes, but not what is projected.

Therefore, though what is projected is not itself included, its possibility is.

A proposition, therefore, does not actually contain its sense, but does contain the possibility of expressing it.

('The content of a proposition' means the content of a proposition that has sense.)

A proposition contains the form, but not the content, of its sense.

3.14 What constitutes a propositional sign is that in it its elements (the words) stand in a determinate relation to one another.

A propositional sign is a fact.

3.141 A proposition is not a blend of words.—(Just as a theme in music is not a blend of notes.)

A proposition is articulate.

3.142 Only facts can express a sense, a set of names cannot.

3.143 Although a propositional sign is a fact, this is obscured by the usual form of expression in writing or print.

For in a printed proposition, for example, no essential difference is apparent between a propositional sign and a word.

(That is what made it possible for Frege to call a proposition a composite name.)

3.1431 The essence of a propositional sign is very clearly seen if we imagine one composed of spatial objects (such as tables, chairs, and books) instead of written signs.

Then the spatial arrangement of these things will express the sense of the proposition.

3.1432 Nicht: „Das komplexe Zeichen ‚aRb' sagt, daß a in der Beziehung R zu b steht", sondern: Daß „a" in einer gewissen Beziehung zu „b" steht, sagt, daß aRb.

3.144 Sachlagen kann man beschreiben, nicht benennen.

(Namen gleichen Punkten, Sätze Pfeilen, sie haben Sinn.)

3.2 Im Satze kann der Gedanke so ausgedrückt sein, daß den Gegenständen des Gedankens Elemente des Satzzeichens entsprechen.

3.201 Diese Elemente nenne ich „einfache Zeichen" und den Satz „vollständig analysiert".

3.202 Die im Satze angewandten einfachen Zeichen heißen Namen.

3.203 Der Name bedeutet den Gegenstand. Der Gegenstand ist seine Bedeutung. („A" ist dasselbe Zeichen wie „A".)

3.21 Der Konfiguration der einfachen Zeichen im Satzzeichen entspricht die Konfiguration der Gegenstände in der Sachlage.

3.22 Der Name vertritt im Satz den Gegenstand.

3.221 Die Gegenstände kann ich nur nennen. Zeichen vertreten sie. Ich kann nur von ihnen sprechen, sie aussprechen kann ich nicht. Ein Satz kann nur sagen, wie ein Ding ist, nicht was es ist.

3.23 Die Forderung der Möglichkeit der einfachen Zeichen ist die Forderung der Bestimmtheit des Sinnes.

3.24 Der Satz, welcher vom Komplex handelt, steht in interner Beziehung zum Satze, der von dessen Bestandteil handelt.

Der Komplex kann nur durch seine Beschreibung gegeben sein, und diese wird stimmen oder nicht stimmen. Der Satz, in welchem von einem Komplex die Rede ist, wird, wenn dieser nicht existiert, nicht unsinnig, sondern einfach falsch sein.

Daß ein Satzelement einen Komplex bezeichnet, kann man aus einer Unbestimmtheit in den Sätzen sehen, worin es vorkommt. Wir wissen, durch diesen Satz ist noch nicht alles bestimmt. (Die Allgemeinheitsbezeichnung enthält ja ein Urbild.)

3.1432 Instead of, 'The complex sign "*a*R*b*" says that *a* stands to *b* in the relation R', we ought to put, '*That* "*a*" stands to "*b*" in a certain relation says *that a*R*b*.'

3.144 Situations can be described but not *given names*.

(Names are like points; propositions like arrows—they have sense.)

3.2 In a proposition a thought can be expressed in such a way that elements of the propositional sign correspond to the objects of the thought.

3.201 I call such elements 'simple signs', and such a proposition 'completely analysed'.

3.202 The simple signs employed in propositions are called names.

3.203 A name means an object. The object is its meaning. ('*A*' is the same sign as '*A*'.)

3.21 The configuration of objects in a situation corresponds to the configuration of simple signs in the propositional sign.

3.22 In a proposition a name is the representative of an object.

3.221 Objects can only be *named*. Signs are their representatives. I can only speak *about* them: I cannot *put them into words*. Propositions can only say *how* things are, not *what* they are.

3.23 The requirement that simple signs be possible is the requirement that sense be determinate.

3.24 A proposition about a complex stands in an internal relation to a proposition about a constituent of the complex.

A complex can be given only by its description, which will be right or wrong. A proposition that mentions a complex will not be nonsensical, if the complex does not exist, but simply false.

When a propositional element signifies a complex, this can be seen from an indeterminateness in the propositions in which it occurs. In such cases we *know* that the proposition leaves something undetermined. (In fact the notation for generality *contains* a prototype.)

Die Zusammenfassung des Symbols eines Komplexes in ein einfaches Symbol kann durch eine Definition ausgedrückt werden.

3.25 Es gibt eine und nur eine vollständige Analyse des Satzes.

3.251 Der Satz drückt auf bestimmte, klar angebbare Weise aus, was er ausdrückt: Der Satz ist artikuliert.

3.26 Der Name ist durch keine Definition weiter zu zergliedern: er ist ein Urzeichen.

3.261 Jedes definierte Zeichen bezeichnet über jene Zeichen, durch welche es definiert wurde; und die Definitionen weisen den Weg.

Zwei Zeichen, ein Urzeichen, und ein durch Urzeichen definiertes, können nicht auf dieselbe Art und Weise bezeichnen. Namen kann man nicht durch Definitionen auseinanderlegen. (Kein Zeichen, welches allein, selbständig eine Bedeutung hat.)

3.262 Was in den Zeichen nicht zum Ausdruck kommt, das zeigt ihre Anwendung. Was die Zeichen verschlucken, das spricht ihre Anwendung aus.

3.263 Die Bedeutungen von Urzeichen können durch Erläuterungen erklärt werden. Erläuterungen sind Sätze, welche die Urzeichen enthalten. Sie können also nur verstanden werden, wenn die Bedeutungen dieser Zeichen bereits bekannt sind.

3.3 Nur der Satz hat Sinn; nur im Zusammenhange des Satzes hat ein Name Bedeutung.

3.31 Jeden Teil des Satzes, der seinen Sinn charakterisiert, nenne ich einen Ausdruck (ein Symbol).

(Der Satz selbst ist ein Ausdruck.)

Ausdruck ist alles, für den Sinn des Satzes wesentliche, was Sätze miteinander gemein haben können.

Der Ausdruck kennzeichnet eine Form und einen Inhalt.

3.311 Der Ausdruck setzt die Formen aller Sätze voraus, in welchen er vorkommen kann. Er ist das gemeinsame charakteristische Merkmal einer Klasse von Sätzen.

The contraction of a symbol for a complex into a simple symbol can be expressed in a definition.

3.25 A proposition has one and only one complete analysis.

3.251 What a proposition expresses it expresses in a determinate manner, which can be set out clearly: a proposition is articulate.

3.26 A name cannot be dissected any further by means of a definition: it is a primitive sign.

3.261 Every sign that has a definition signifies *via* the signs that serve to define it; and the definitions point the way.

Two signs cannot signify in the same manner if one is primitive and the other is defined by means of primitive signs. Names *cannot* be anatomized by means of definitions. (Nor can any sign that has a meaning independently and on its own.)

3.262 What signs fail to express, their application shows. What signs slur over, their application says clearly.

3.263 The meanings of primitive signs can be explained by means of elucidations. Elucidations are propositions that contain the primitive signs. So they can only be understood if the meanings of those signs are already known.

3.3 Only propositions have sense; only in the nexus of a proposition does a name have meaning.

3.31 I call any part of a proposition that characterizes its sense an expression (or a symbol).

(A proposition is itself an expression.)

Everything essential to their sense that propositions can have in common with one another is an expression.

An expression is the mark of a form and a content.

3.311 An expression presupposes the forms of all the propositions in which it can occur. It is the common characteristic mark of a class of propositions.

3.312 Er wird also dargestellt durch die allgemeine Form der Sätze, die er charakterisiert.

Und zwar wird in dieser Form der Ausdruck *konstant* und alles übrige *variabel* sein.

3.313 Der Ausdruck wird also durch eine Variable dargestellt, deren Werte die Sätze sind, die den Ausdruck enthalten.

(Im Grenzfall wird die Variable zur Konstanten, der Ausdruck zum Satz.)

Ich nenne eine solche Variable „Satzvariable".

3.314 Der Ausdruck hat nur im Satz Bedeutung. Jede Variable läßt sich als Satzvariable auffassen.

(Auch der variable Name.)

3.315 Verwandeln wir einen Bestandteil eines Satzes in eine Variable, so gibt es eine Klasse von Sätzen, welche sämtlich Werte des so entstandenen variablen Satzes sind. Diese Klasse hängt im allgemeinen noch davon ab, was wir, nach willkürlicher Übereinkunft, mit Teilen jenes Satzes meinen. Verwandeln wir aber alle jene Zeichen, deren Bedeutung willkürlich bestimmt wurde, in Variable, so gibt es nun noch immer eine solche Klasse. Diese aber ist nun von keiner Übereinkunft abhängig, sondern nur noch von der Natur des Satzes. Sie entspricht einer logischen Form — einem logischen Urbild.

3.316 Welche Werte die Satzvariable annehmen darf, wird festgesetzt.

Die Festsetzung der Werte *ist* die Variable.

3.317 Die Festsetzung der Werte der Satzvariablen ist die *Angabe der Sätze*, deren gemeinsames Merkmal die Variable ist.

Die Festsetzung ist eine Beschreibung dieser Sätze.

Die Festsetzung wird also nur von Symbolen, nicht von deren Bedeutung handeln.

Und *nur* dies ist der Festsetzung wesentlich, *daß sie nur eine Beschreibung von Symbolen ist und nichts über das Bezeichnete aussagt.*

Wie die Beschreibung der Sätze geschieht, ist unwesentlich.

3.318 Den Satz fasse ich — wie Frege und Russell — als Funktion der in ihm enthaltenen Ausdrücke auf.

3.312 It is therefore presented by means of the general form of the propositions that it characterizes.

In fact, in this form the expression will be *constant* and everything else *variable*.

3.313 Thus an expression is presented by means of a variable whose values are the propositions that contain the expression.

(In the limiting case the variable becomes a constant, the expression becomes a proposition.)

I call such a variable a 'propositional variable'.

3.314 An expression has meaning only in a proposition. All variables can be construed as propositional variables.

(Even variable names.)

3.315 If we turn a constituent of a proposition into a variable, there is a class of propositions all of which are values of the resulting variable proposition. In general, this class too will be dependent on the meaning that our arbitrary conventions have given to parts of the original proposition. But if all the signs in it that have arbitrarily determined meanings are turned into variables, we shall still get a class of this kind. This one, however, is not dependent on any convention, but solely on the nature of the proposition. It corresponds to a logical form—a logical prototype.

3.316 What values a propositional variable may take is something that is stipulated.

The stipulation of values *is* the variable.

3.317 To stipulate values for a propositional variable is *to give the propositions* whose common characteristic the variable is.

The stipulation is a description of those propositions.

The stipulation will therefore be concerned only with symbols, not with their meaning.

And the *only* thing essential to the stipulation is *that it is merely a description of symbols and states nothing about what is signified.*

How the description of the propositions is produced is not essential.

3.318 Like Frege and Russell I construe a proposition as a function of the expressions contained in it.

3.32 Das Zeichen ist das sinnlich Wahrnehmbare am Symbol.

3.321 Zwei verschiedene Symbole können also das Zeichen (Schriftzeichen oder Lautzeichen etc.) miteinander gemein haben — sie bezeichnen dann auf verschiedene Art und Weise.

3.322 Es kann nie das gemeinsame Merkmal zweier Gegenstände anzeigen, daß wir sie mit demselben Zeichen, aber durch zwei verschiedene B e z e i c h n u n g s w e i s e n bezeichnen. Denn das Zeichen ist ja willkürlich. Man könnte also auch zwei verschiedene Zeichen wählen, und wo bliebe dann das Gemeinsame in der Bezeichnung?

3.323 In der Umgangssprache kommt es ungemein häufig vor, daß dasselbe Wort auf verschiedene Art und Weise bezeichnet — also verschiedenen Symbolen angehört —, oder, daß zwei Wörter, die auf verschiedene Art und Weise bezeichnen, äußerlich in der gleichen Weise im Satze angewandt werden.

So erscheint das Wort „ist" als Kopula, als Gleichheitszeichen und als Ausdruck der Existenz; „existieren" als intransitives Zeitwort wie „gehen"; „identisch" als Eigenschaftswort; wir reden von E t w a s, aber auch davon, daß e t w a s geschieht.

(Im Satze: „Grün ist grün"— wo das erste Wort ein Personenname, das letzte ein Eigenschaftswort ist — haben diese Worte nicht einfach verschiedene Bedeutung, sondern es sind v e r s c h i e d e n e S y m b o l e.)

3.324 So entstehen leicht die fundamentalsten Verwechslungen (deren die ganze Philosophie voll ist).

3.325 Um diesen Irrtümern zu entgehen, müssen wir eine Zeichensprache verwenden, welche sie ausschließt, indem sie nicht das gleiche Zeichen in verschiedenen Symbolen, und Zeichen, welche auf verschiedene Art bezeichnen, nicht äußerlich auf die gleiche Art verwendet. Eine Zeichensprache also, die der l o g i s c h e n Grammatik — der logischen Syntax — gehorcht.

(Die Begriffsschrift Freges und Russells ist eine solche Sprache, die allerdings noch nicht alle Fehler ausschließt.)

3.32 A sign is what can be perceived of a symbol.

3.321 So one and the same sign (written or spoken, etc.) can be common to two different symbols—in which case they will signify in different ways.

3.322 Our use of the same sign to signify two different objects can never indicate a common characteristic of the two, if we use it with two different *modes of signification*. For the sign, of course, is arbitrary. So we could choose two different signs instead, and then what would be left in common on the signifying side?

3.323 In everyday language it very frequently happens that the same word has different modes of signification—and so belongs to different symbols—or that two words that have different modes of signification are employed in propositions in what is superficially the same way.

Thus the word 'is' figures as the copula, as a sign for identity, and as an expression for existence; 'exist' figures as an intransitive verb like 'go', and 'identical' as an adjective; we speak of *something*, but also of *something's* happening.

(In the proposition, 'Green is green'—where the first word is the proper name of a person and the last an adjective—these words do not merely have different meanings: they are *different symbols*.)

3.324 In this way the most fundamental confusions are easily produced (the whole of philosophy is full of them).

3.325 In order to avoid such errors we must make use of a sign-language that excludes them by not using the same sign for different symbols and by not using in a superficially similar way signs that have different modes of signification: that is to say, a sign-language that is governed by *logical* grammar—by logical syntax.

(The conceptual notation of Frege and Russell is such a language, though, it is true, it fails to exclude all mistakes.)

3.326 Um das Symbol am Zeichen zu erkennen, muß man auf den sinnvollen Gebrauch achten.

3.327 Das Zeichen bestimmt erst mit seiner logisch-syntaktischen Verwendung zusammen eine logische Form.

3.328 Wird ein Zeichen *nicht gebraucht*, so ist es bedeutungslos. Das ist der Sinn der Devise Occams.

(Wenn sich alles so verhält als hätte ein Zeichen Bedeutung, dann hat es auch Bedeutung.)

3.33 In der logischen Syntax darf nie die Bedeutung eines Zeichens eine Rolle spielen; sie muß sich aufstellen lassen, ohne daß dabei von der *Bedeutung* eines Zeichens die Rede wäre, sie darf *nur* die Beschreibung der Ausdrücke voraussetzen.

3.331 Von dieser Bemerkung sehen wir in Russells „Theory of Types" hinüber: Der Irrtum Russells zeigt sich darin, daß er bei der Aufstellung der Zeichenregeln von der Bedeutung der Zeichen reden mußte.

3.332 Kein Satz kann etwas über sich selbst aussagen, weil das Satzzeichen nicht in sich selbst enthalten sein kann (das ist die ganze „Theory of Types").

3.333 Eine Funktion kann darum nicht ihr eigenes Argument sein, weil das Funktionszeichen bereits das Urbild seines Arguments enthält und es sich nicht selbst enthalten kann.

Nehmen wir nämlich an, die Funktion F(fx) könnte ihr eigenes Argument sein; dann gäbe es also einen Satz: „F(F(fx))", und in diesem müssen die äußere Funktion F und die innere Funktion F verschiedene Bedeutungen haben, denn die innere hat die Form ϕ(fx), die äußere, die Form ψ(ϕ(fx)). Gemeinsam ist den beiden Funktionen nur der Buchstabe „F", der aber allein nichts bezeichnet.

Dies wird sofort klar, wenn wir statt „F(Fu)" schreiben „($\exists\phi$):F(ϕu).ϕu = Fu".

Hiermit erledigt sich Russells Paradox.

3.334 Die Regeln der logischen Syntax müssen sich von selbst verstehen, wenn man nur weiß, wie ein jedes Zeichen bezeichnet.

3.326 In order to recognize a symbol by its sign we must observe how it is used with a sense.

3.327 A sign does not determine a logical form unless it is taken together with its logico-syntactical employment.

3.328 If a sign is *useless*, it is meaningless. That is the point of Occam's maxim.

(If everything behaves as if a sign had meaning, then it does have meaning.)

3.33 In logical syntax the meaning of a sign should never play a rôle. It must be possible to establish logical syntax without mentioning the *meaning* of a sign: *only* the description of expressions may be presupposed.

3.331 From this observation we turn to Russell's 'theory of types'. It can be seen that Russell must be wrong, because he had to mention the meaning of signs when establishing the rules for them.

3.332 No proposition can make a statement about itself, because a propositional sign cannot be contained in itself (that is the whole of the 'theory of types').

3.333 The reason why a function cannot be its own argument is that the sign for a function already contains the prototype of its argument, and it cannot contain itself.

For let us suppose that the function $F(fx)$ could be its own argument: in that case there would be a proposition '$F(F(fx))$', in which the outer function F and the inner function F must have different meanings, since the inner one has the form $\phi(fx)$ and the outer one has the form $\psi(\phi(fx))$. Only the letter 'F' is common to the two functions, but the letter by itself signifies nothing.

This immediately becomes clear if instead of '$F(Fu)$' we write '$(\exists\phi):F(\phi u).\phi u = Fu$'.

That disposes of Russell's paradox.

3.334 The rules of logical syntax must go without saying, once we know how each individual sign signifies.

3.34 Der Satz besitzt wesentliche und zufällige Züge.

Zufällig sind die Züge, die von der besonderen Art der Hervorbringung des Satzzeichens herrühren. Wesentlich diejenigen, welche allein den Satz befähigen, seinen Sinn auszudrücken.

3.341 Das Wesentliche am Satz ist also das, was allen Sätzen, welche den gleichen Sinn ausdrücken können, gemeinsam ist.

Und ebenso ist allgemein das Wesentliche am Symbol das, was alle Symbole, die denselben Zweck erfüllen können, gemeinsam haben.

3.3411 Man könnte also sagen: Der eigentliche Name ist das, was alle Symbole, die den Gegenstand bezeichnen, gemeinsam haben. Es würde sich so successive ergeben, daß keinerlei Zusammensetzung für den Namen wesentlich ist.

3.342 An unseren Notationen ist zwar etwas willkürlich, aber *das* ist nicht willkürlich: Daß, *wenn* wir etwas willkürlich bestimmt haben, dann etwas anderes der Fall sein muß. (Dies hängt von dem *Wesen* der Notation ab.)

3.3421 Eine besondere Bezeichnungsweise mag unwichtig sein, aber wichtig ist es immer, daß diese eine *mögliche* Bezeichnungsweise ist. Und so verhält es sich in der Philosophie überhaupt: Das Einzelne erweist sich immer wieder als unwichtig, aber die Möglichkeit jedes Einzelnen gibt uns einen Aufschluß über das Wesen der Welt.

3.343 Definitionen sind Regeln der Übersetzung von einer Sprache in eine andere. Jede richtige Zeichensprache muß sich in jede andere nach solchen Regeln übersetzen lassen: *Dies* ist, was sie alle gemeinsam haben.

3.344 Das, was am Symbol bezeichnet, ist das Gemeinsame aller jener Symbole, durch die das erste den Regeln der logischen Syntax zufolge ersetzt werden kann.

3.3441 Man kann z. B. das Gemeinsame aller Notationen für die Wahrheitsfunktionen so ausdrücken: Es ist ihnen gemeinsam, daß sich alle — z. B.— durch die Notation von „~p“ („nicht p“) und „p v q“ („p oder q“) *ersetzen lassen*.

3.34 A proposition possesses essential and accidental features.

Accidental features are those that result from the particular way in which the propositional sign is produced. Essential features are those without which the proposition could not express its sense.

3.341 So what is essential in a proposition is what all propositions that can express the same sense have in common.

And similarly, in general, what is essential in a symbol is what all symbols that can serve the same purpose have in common.

3.3411 So one could say that the real name of an object was what all symbols that signified it had in common. Thus, one by one, all kinds of composition would prove to be unessential to a name.

3.342 Although there is something arbitrary in our notations, *this* much is not arbitrary—that *when* we have determined one thing arbitrarily, something else is necessarily the case. (This derives from the *essence* of notation.)

3.3421 A particular mode of signifying may be unimportant but it is always important that it is a *possible* mode of signifying. And that is generally so in philosophy: again and again the individual case turns out to be unimportant, but the possibility of each individual case discloses something about the essence of the world.

3.343 Definitions are rules for translating from one language into another. Any correct sign-language must be translatable into any other in accordance with such rules: it is *this* that they all have in common.

3.344 What signifies in a symbol is what is common to all the symbols that the rules of logical syntax allow us to substitute for it.

3.3441 For instance, we can express what is common to all notations for truth-functions in the following way: they have in common that, for example, the notation that uses '$\sim p$' ('not p') and '$p \vee q$' ('p or q') *can be substituted* for any of them.

(Hiermit ist die Art und Weise gekennzeichnet, wie eine spezielle mögliche Notation uns allgemeine Aufschlüsse geben kann.)

3.3442 Das Zeichen des Komplexes löst sich auch bei der Analyse nicht willkürlich auf, so daß etwa seine Auflösung in jedem Satzgefüge eine andere wäre.

3.4 Der Satz bestimmt einen Ort im logischen Raum. Die Existenz dieses logischen Ortes ist durch die Existenz der Bestandteile allein verbürgt, durch die Existenz des sinnvollen Satzes.

3.41 Das Satzzeichen und die logischen Koordinaten: Das ist der logische Ort.

3.411 Der geometrische und der logische Ort stimmen darin überein, daß beide die Möglichkeit einer Existenz sind.

3.42 Obwohl der Satz nur einen Ort des logischen Raumes bestimmen darf, so muß doch durch ihn schon der ganze logische Raum gegeben sein.

(Sonst würden durch die Verneinung, die logische Summe, das logische Produkt, etc. immer neue Elemente — in Koordination — eingeführt.)

(Das logische Gerüst um das Bild herum bestimmt den logischen Raum. Der Satz durchgreift den ganzen logischen Raum.)

3.5 Das angewandte, gedachte, Satzzeichen ist der Gedanke.

4 Der Gedanke ist der sinnvolle Satz.

4.001 Die Gesamtheit der Sätze ist die Sprache.

4.002 Der Mensch besitzt die Fähigkeit Sprachen zu bauen, womit sich jeder Sinn ausdrücken läßt, ohne eine Ahnung davon zu haben, wie und was jedes Wort bedeutet.— Wie man auch spricht, ohne zu wissen, wie die einzelnen Laute hervorgebracht werden.

Die Umgangssprache ist ein Teil des menschlichen Organismus und nicht weniger kompliziert als dieser.

Es ist menschenunmöglich, die Sprachlogik aus ihr unmittelbar zu entnehmen.

(This serves to characterize the way in which something general can be disclosed by the possibility of a specific notation.)

3.3442 Nor does analysis resolve the sign for a complex in an arbitrary way, so that it would not have a different resolution every time that it was incorporated in a different proposition.

3.4 A proposition determines a place in logical space. The existence of this logical place is guaranteed by the mere existence of the constituents—by the existence of the proposition with a sense.

3.41 The propositional sign with logical co-ordinates—that is the logical place.

3.411 In geometry and logic alike a place is a possibility: something can exist in it.

3.42 A proposition can determine only one place in logical space: nevertheless the whole of logical space must already be given by it.

(Otherwise negation, logical sum, logical product, etc., would introduce more and more new elements—in co-ordination.)

(The logical scaffolding surrounding a picture determines logical space. The force of a proposition reaches through the whole of logical space.)

3.5 A propositional sign, applied and thought out, is a thought.

4 A thought is a proposition with a sense.

4.001 The totality of propositions is language.

4.002 Man possesses the ability to construct languages capable of expressing every sense, without having any idea how each word has meaning or what its meaning is—just as people speak without knowing how the individual sounds are produced.

Everyday language is a part of the human organism and is no less complicated than it.

It is not humanly possible to gather immediately from it what the logic of language is.

Die Sprache verkleidet den Gedanken. Und zwar so, daß man nach der äußeren Form des Kleides nicht auf die Form des bekleideten Gedankens schließen kann; weil die äußere Form des Kleides nach ganz anderen Zwecken gebildet ist, als danach, die Form des Körpers erkennen zu lassen.

Die stillschweigenden Abmachungen zum Verständnis der Umgangssprache sind enorm kompliziert.

4.003 Die meisten Sätze und Fragen, welche über philosophische Dinge geschrieben worden sind, sind nicht falsch, sondern unsinnig. Wir können daher Fragen dieser Art überhaupt nicht beantworten, sondern nur ihre Unsinnigkeit feststellen. Die meisten Fragen und Sätze der Philosophen beruhen darauf, daß wir unsere Sprachlogik nicht verstehen.

(Sie sind von der Art der Frage, ob das Gute mehr oder weniger identisch sei als das Schöne.)

Und es ist nicht verwunderlich, daß die tiefsten Probleme eigentlich k e i n e Probleme sind.

4.0031 Alle Philosophie ist „Sprachkritik". (Allerdings nicht im Sinne Mauthners.) Russells Verdienst ist es, gezeigt zu haben, daß die scheinbare logische Form des Satzes nicht seine wirkliche sein muß.

4.01 Der Satz ist ein Bild der Wirklichkeit.

Der Satz ist ein Modell der Wirklichkeit, so wie wir sie uns denken.

4.011 Auf den ersten Blick scheint der Satz — wie er etwa auf dem Papier gedruckt steht — kein Bild der Wirklichkeit zu sein, von der er handelt. Aber auch die Notenschrift scheint auf den ersten Blick kein Bild der Musik zu sein, und unsere Lautzeichen- (Buchstaben-) Schrift kein Bild unserer Lautsprache.

Und doch erweisen sich diese Zeichensprachen auch im gewöhnlichen Sinne als Bilder dessen, was sie darstellen.

4.012 Offenbar ist, daß wir einen Satz von der Form „aRb" als Bild empfinden. Hier ist das Zeichen offenbar ein Gleichnis des Bezeichneten.

Language disguises thought. So much so, that from the outward form of the clothing it is impossible to infer the form of the thought beneath it, because the outward form of the clothing is not designed to reveal the form of the body, but for entirely different purposes.

The tacit conventions on which the understanding of everyday language depends are enormously complicated.

4.003 Most of the propositions and questions to be found in philosophical works are not false but nonsensical. Consequently we cannot give any answer to questions of this kind, but can only point out that they are nonsensical. Most of the propositions and questions of philosophers arise from our failure to understand the logic of our language.

(They belong to the same class as the question whether the good is more or less identical than the beautiful.)

And it is not surprising that the deepest problems are in fact *not* problems at all.

4.0031 All philosophy is a 'critique of language' (though not in Mauthner's sense). It was Russell who performed the service of showing that the apparent logical form of a proposition need not be its real one.

4.01 A proposition is a picture of reality.

A proposition is a model of reality as we imagine it.

4.011 At first sight a proposition—one set out on the printed page, for example—does not seem to be a picture of the reality with which it is concerned. But neither do written notes seem at first sight to be a picture of a piece of music, nor our phonetic notation (the alphabet) to be a picture of our speech.

And yet these sign-languages prove to be pictures, even in the ordinary sense, of what they represent.

4.012 It is obvious that a proposition of the form '*a*R*b*' strikes us as a picture. In this case the sign is obviously a likeness of what is signified.

4.013 Und wenn wir in das Wesentliche dieser Bildhaftigkeit eindringen, so sehen wir, daß dieselbe durch *scheinbare Unregelmäßigkeiten* (wie die Verwendung der ♯ und ♭ in der Notenschrift) *nicht* gestört wird.

Denn auch diese Unregelmäßigkeiten bilden das ab, was sie ausdrücken sollen; nur auf eine andere Art und Weise.

4.014 Die Grammophonplatte, der musikalische Gedanke, die Notenschrift, die Schallwellen, stehen alle in jener abbildenden internen Beziehung zu einander, die zwischen Sprache und Welt besteht.

Ihnen allen ist der logische Bau gemeinsam.

(Wie im Märchen die zwei Jünglinge, ihre zwei Pferde und ihre Lilien. Sie sind alle in gewissem Sinne Eins.)

4.0141 Daß es eine allgemeine Regel gibt, durch die der Musiker aus der Partitur die Symphonie entnehmen kann, durch welche man aus der Linie auf der Grammophonplatte die Symphonie und nach der ersten Regel wieder die Partitur ableiten kann, darin besteht eben die innere Ähnlichkeit dieser scheinbar so ganz verschiedenen Gebilde. Und jene Regel ist das Gesetz der Projektion, welches die Symphonie in die Notensprache projiziert. Sie ist die Regel der Übersetzung der Notensprache in die Sprache der Grammophonplatte.

4.015 Die Möglichkeit aller Gleichnisse, der ganzen Bildhaftigkeit unserer Ausdrucksweise, ruht in der Logik der Abbildung.

4.016 Um das Wesen des Satzes zu verstehen, denken wir an die Hieroglyphenschrift, welche die Tatsachen die sie beschreibt abbildet.

Und aus ihr wurde die Buchstabenschrift, ohne das Wesentliche der Abbildung zu verlieren.

4.02 Dies sehen wir daraus, daß wir den Sinn des Satzzeichens verstehen, ohne daß er uns erklärt wurde.

4.021 Der Satz ist ein Bild der Wirklichkeit: Denn ich kenne die von ihm dargestellte Sachlage, wenn ich den Satz

4.013 And if we penetrate to the essence of this pictorial character, we see that it is *not* impaired by *apparent irregularities* (such as the use of ♯ and ♭ in musical notation).

For even these irregularities depict what they are intended to express; only they do it in a different way.

4.014 A gramophone record, the musical idea, the written notes, and the sound-waves, all stand to one another in the same internal relation of depicting that holds between language and the world.

They are all constructed according to a common logical pattern.

(Like the two youths in the fairy-tale, their two horses, and their lilies. They are all in a certain sense one.)

4.0141 There is a general rule by means of which the musician can obtain the symphony from the score, and which makes it possible to derive the symphony from the groove on the gramophone record, and, using the first rule, to derive the score again. That is what constitutes the inner similarity between these things which seem to be constructed in such entirely different ways. And that rule is the law of projection which projects the symphony into the language of musical notation. It is the rule for translating this language into the language of gramophone records.

4.015 The possibility of all imagery, of all our pictorial modes of expression, is contained in the logic of depiction.

4.016 In order to understand the essential nature of a proposition, we should consider hieroglyphic script, which depicts the facts that it describes.

And alphabetic script developed out of it without losing what was essential to depiction.

4.02 We can see this from the fact that we understand the sense of a propositional sign without its having been explained to us.

4.021 A proposition is a picture of reality: for if I understand a proposition, I know the situation that it represents. And

verstehe. Und den Satz verstehe ich, ohne daß mir sein Sinn erklärt wurde.

4.022 Der Satz *zeigt* seinen Sinn.

Der Satz *zeigt*, wie es sich verhält, *wenn* er wahr ist. Und er *sagt*, *daß* es sich so verhält.

4.023 Die Wirklichkeit muß durch den Satz auf ja oder nein fixiert sein.

Dazu muß sie durch ihn vollständig beschrieben werden.

Der Satz ist die Beschreibung eines Sachverhaltes.

Wie die Beschreibung einen Gegenstand nach seinen externen Eigenschaften, so beschreibt der Satz die Wirklichkeit nach ihren internen Eigenschaften.

Der Satz konstruiert eine Welt mit Hilfe eines logischen Gerüstes und darum kann man am Satz auch sehen, wie sich alles Logische verhält, *wenn* er wahr ist. Man kann aus einem falschen Satz *Schlüsse ziehen*.

4.024 Einen Satz verstehen, heißt, wissen was der Fall ist, wenn er wahr ist.

(Man kann ihn also verstehen, ohne zu wissen, ob er wahr ist.)

Man versteht ihn, wenn man seine Bestandteile versteht.

4.025 Die Übersetzung einer Sprache in eine andere geht nicht so vor sich, daß man jeden *Satz* der einen in einen *Satz* der anderen übersetzt, sondern nur die Satzbestandteile werden übersetzt.

(Und das Wörterbuch übersetzt nicht nur Substantiva, sondern auch Zeit-, Eigenschafts- und Bindewörter etc.; und es behandelt sie alle gleich.)

4.026 Die Bedeutungen der einfachen Zeichen (der Wörter) müssen uns erklärt werden, daß wir sie verstehen.

Mit den Sätzen aber verständigen wir uns.

4.027 Es liegt im Wesen des Satzes, daß er uns einen *neuen* Sinn mitteilen kann.

4.03 Ein Satz muß mit alten Ausdrücken einen neuen Sinn mitteilen.

I understand the proposition without having had its sense explained to me.

4.022 A proposition *shows* its sense.

A proposition *shows* how things stand *if* it is true. And it *says that* they do so stand.

4.023 A proposition must restrict reality to two alternatives: yes or no.

In order to do that, it must describe reality completely.

A proposition is a description of a state of affairs.

Just as a description of an object describes it by giving its external properties, so a proposition describes reality by its internal properties.

A proposition constructs a world with the help of a logical scaffolding, so that one can actually see from the proposition how everything stands logically *if* it is true. One can *draw inferences* from a false proposition.

4.024 To understand a proposition means to know what is the case if it is true.

(One can understand it, therefore, without knowing whether it is true.)

It is understood by anyone who understands its constituents.

4.025 When translating one language into another, we do not proceed by translating each *proposition* of the one into a *proposition* of the other, but merely by translating the constituents of propositions.

(And the dictionary translates not only substantives, but also verbs, adjectives, and conjunctions, etc.; and it treats them all in the same way.)

4.026 The meanings of simple signs (words) must be explained to us if we are to understand them.

With propositions, however, we make ourselves understood.

4.027 It belongs to the essence of a proposition that it should be able to communicate a *new* sense to us.

4.03 A proposition must use old expressions to communicate a new sense.

Der Satz teilt uns eine Sachlage mit, also muß er *wesentlich* mit der Sachlage zusammenhängen.

Und der Zusammenhang ist eben, daß er ihr logisches Bild ist.

Der Satz sagt nur insoweit etwas aus, als er ein Bild ist.

4.031 Im Satz wird gleichsam eine Sachlage probeweise zusammengestellt.

Man kann geradezu sagen — statt: Dieser Satz hat diesen und diesen Sinn —: Dieser Satz stellt diese und diese Sachlage dar.

4.0311 Ein Name steht für ein Ding, ein anderer für ein anderes Ding und untereinander sind sie verbunden, so stellt das Ganze — wie ein lebendes Bild — den Sachverhalt vor.

4.0312 Die Möglichkeit des Satzes beruht auf dem Prinzip der Vertretung von Gegenständen durch Zeichen.

Mein Grundgedanke ist, daß die „logischen Konstanten“ nicht vertreten. Daß sich die *Logik* der Tatsachen nicht vertreten läßt.

4.032 Nur insoweit ist der Satz ein Bild einer Sachlage, als er logisch gegliedert ist.

(Auch der Satz: „Ambulo“, ist zusammengesetzt, denn sein Stamm ergibt mit einer anderen Endung, und seine Endung mit einem anderen Stamm, einen anderen Sinn.)

4.04 Am Satz muß gerade soviel zu unterscheiden sein, als an der Sachlage, die er darstellt.

Die beiden müssen die gleiche logische (mathematische) Mannigfaltigkeit besitzen. (Vergleiche Hertz' „Mechanik“, über dynamische Modelle.)

4.041 Diese mathematische Mannigfaltigkeit kann man natürlich nicht selbst wieder abbilden. Aus ihr kann man beim Abbilden nicht heraus.

4.0411 Wollten wir z. B. das, was wir durch „(x).fx“ ausdrücken, durch Vorsetzen eines Indexes vor „fx“ ausdrücken — etwa so: „Alg. fx“ — es würde nicht genügen — wir wüßten nicht, was verallgemeinert wurde. Wollten wir es durch einen Index „$_\alpha$“ anzeigen — etwa so: „$f(x_\alpha)$“ — es würde auch nicht genügen — wir wüßten nicht den Bereich der Allgemeinheitsbezeichnung.

A proposition communicates a situation to us, and so it must be *essentially* connected with the situation.

And the connexion is precisely that it is its logical picture.

A proposition states something only in so far as it is a picture.

4.031 In a proposition a situation is, as it were, constructed by way of experiment.

Instead of, 'This proposition has such and such a sense', we can simply say, 'This proposition represents such and such a situation'.

4.0311 One name stands for one thing, another for another thing, and they are combined with one another. In this way the whole group—like a tableau vivant—presents a state of affairs.

4.0312 The possibility of propositions is based on the principle that objects have signs as their representatives.

My fundamental idea is that the 'logical constants' are not representatives; that there can be no representatives of the *logic* of facts.

4.032 It is only in so far as a proposition is logically articulated that it is a picture of a situation.

(Even the proposition, 'Ambulo', is composite: for its stem with a different ending yields a different sense, and so does its ending with a different stem.)

4.04 In a proposition there must be exactly as many distinguishable parts as in the situation that it represents.

The two must possess the same logical (mathematical) multiplicity. (Compare Hertz's *Mechanics* on dynamical models.)

4.041 This mathematical multiplicity, of course, cannot itself be the subject of depiction. One cannot get away from it when depicting.

4.0411 If, for example, we wanted to express what we now write as '$(x).fx$' by putting an affix in front of 'fx'—for instance by writing 'Gen. fx'—it would not be adequate: we should not know what was being generalized. If we wanted to signalize it with an affix '$_g$'—for instance by writing '$f(x_g)$'—that would not be adequate either: we should not know the scope of the generality-sign.

Wollten wir es durch Einführung einer Marke in die Argumentstellen versuchen — etwa so:

„(A, A).F (A, A)"

— es würde nicht genügen — wir könnten die Identität der Variablen nicht feststellen. U.s.w.

Alle diese Bezeichnungsweisen genügen nicht, weil sie nicht die notwendige mathematische Mannigfaltigkeit haben.

4.0412 Aus demselben Grunde genügt die idealistische Erklärung des Sehens der räumlichen Beziehungen durch die „Raumbrille" nicht, weil sie nicht die Mannigfaltigkeit dieser Beziehungen erklären kann.

4.05 Die Wirklichkeit wird mit dem Satz verglichen.

4.06 Nur dadurch kann der Satz wahr oder falsch sein, indem er ein Bild der Wirklichkeit ist.

4.061 Beachtet man nicht, daß der Satz einen von den Tatsachen unabhängigen Sinn hat, so kann man leicht glauben, daß wahr und falsch gleichberechtigte Beziehungen von Zeichen und Bezeichnetem sind.

Man könnte dann z. B. sagen, daß „p" auf die wahre Art bezeichnet, was „∼p" auf die falsche Art, etc.

4.062 Kann man sich nicht mit falschen Sätzen, wie bisher mit wahren, verständigen? Solange man nur weiß, daß sie falsch gemeint sind. Nein! Denn, wahr ist ein Satz, wenn es sich so verhält, wie wir es durch ihn sagen; und wenn wir mit „p" ∼ p meinen, und es sich so verhält wie wir es meinen, so ist „p" in der neuen Auffassung wahr und nicht falsch.

4.0621 Daß aber die Zeichen „p" und „∼p" das gleiche sagen k ö n n e n, ist wichtig. Denn es zeigt, daß dem Zeichen „∼" in der Wirklichkeit nichts entspricht.

Daß in einem Satz die Verneinung vorkommt, ist noch kein Merkmal seines Sinnes (∼∼ p = p).

Die Sätze „p" und „∼p" haben entgegengesetzten Sinn, aber es entspricht ihnen eine und dieselbe Wirklichkeit.

4.063 Ein Bild zur Erklärung des Wahrheitsbegriffes:

If we were to try to do it by introducing a mark into the argument-places—for instance by writing

$$`(G,G).F(G,G)'$$

—it would not be adequate: we should not be able to establish the identity of the variables. And so on.

All these modes of signifying are inadequate because they lack the necessary mathematical multiplicity.

4.0412 For the same reason the idealist's appeal to 'spatial spectacles' is inadequate to explain the seeing of spatial relations, because it cannot explain the multiplicity of these relations.

4.05 Reality is compared with propositions.

4.06 A proposition can be true or false only in virtue of being a picture of reality.

4.061 It must not be overlooked that a proposition has a sense that is independent of the facts: otherwise one can easily suppose that true and false are relations of equal status between signs and what they signify.

In that case one could say, for example, that '*p*' signified in the true way what '$\sim p$' signified in the false way, etc.

4.062 Can we not make ourselves understood with false propositions just as we have done up till now with true ones? —So long as it is known that they are meant to be false.— No! For a proposition is true if we use it to say that things stand in a certain way, and they do; and if by '*p*' we mean $\sim p$ and things stand as we mean that they do, then, construed in the new way, '*p*' is true and not false.

4.0621 But it is important that the signs '*p*' and '$\sim p$' *can* say the same thing. For it shows that nothing in reality corresponds to the sign '$\sim$'.

The occurrence of negation in a proposition is not enough to characterize its sense ($\sim\sim p = p$).

The propositions '*p*' and '$\sim p$' have opposite sense, but there corresponds to them one and the same reality.

4.063 An analogy to illustrate the concept of truth: imagine

Schwarzer Fleck auf weißem Papier; die Form des Fleckes kann man beschreiben, indem man für jeden Punkt der Fläche angibt, ob er weiß oder schwarz ist. Der Tatsache, daß ein Punkt schwarz ist, entspricht eine positive — der, daß ein Punkt weiß (nicht schwarz) ist, eine negative Tatsache. Bezeichne ich einen Punkt der Fläche (einen Fregeschen Wahrheitswert), so entspricht dies der Annahme, die zur Beurteilung aufgestellt wird, etc. etc.

Um aber sagen zu können, ein Punkt sei schwarz oder weiß, muß ich vorerst wissen, wann man einen Punkt schwarz und wann man ihn weiß nennt; um sagen zu können: „p" ist wahr (oder falsch), muß ich bestimmt haben, unter welchen Umständen ich „p" wahr nenne, und damit bestimme ich den Sinn des Satzes.

Der Punkt, an dem das Gleichnis hinkt, ist nun der: Wir können auf einen Punkt des Papiers zeigen, auch ohne zu wissen, was weiß und schwarz ist; einem Satz ohne Sinn aber entspricht gar nichts, denn er bezeichnet kein Ding (Wahrheitswert), dessen Eigenschaften etwa „falsch" oder „wahr" hießen; das Verbum eines Satzes ist nicht „ist wahr" oder „ist falsch"— wie Frege glaubte —, sondern das, was „wahr ist", muß das Verbum schon enthalten.

4.064 Jeder Satz muß s c h o n einen Sinn haben; die Bejahung kann ihn ihm nicht geben, denn sie bejaht ja gerade den Sinn. Und dasselbe gilt von der Verneinung, etc.

4.0641 Man könnte sagen: Die Verneinung bezieht sich schon auf den logischen Ort, den der verneinte Satz bestimmt.

Der verneinende Satz bestimmt einen a n d e r e n logischen Ort als der verneinte.

Der verneinende Satz bestimmt einen logischen Ort mit Hilfe des logischen Ortes des verneinten Satzes, indem er jenen als außerhalb diesem liegend beschreibt.

Daß man den verneinten Satz wieder verneinen kann, zeigt schon, daß das, was verneint wird, schon ein Satz und nicht erst die Vorbereitung zu einem Satze ist.

4.1 Der Satz stellt das Bestehen und Nichtbestehen der Sachverhalte dar.

a black spot on white paper: you can describe the shape of the spot by saying, for each point on the sheet, whether it is black or white. To the fact that a point is black there corresponds a positive fact, and to the fact that a point is white (not black), a negative fact. If I designate a point on the sheet (a truth-value according to Frege), then this corresponds to the supposition that is put forward for judgement, etc. etc.

But in order to be able to say that a point is black or white, I must first know when a point is called black, and when white: in order to be able to say, '"*p*" is true (or false)', I must have determined in what circumstances I call '*p*' true, and in so doing I determine the sense of the proposition.

Now the point where the simile breaks down is this: we can indicate a point on the paper even if we do not know what black and white are, but if a proposition has no sense, nothing corresponds to it, since it does not designate a thing (a truth-value) which might have properties called 'false' or 'true'. The verb of a proposition is not 'is true' or 'is false', as Frege thought: rather, that which 'is true' must already contain the verb.

4.064 Every proposition must *already* have a sense: it cannot be given a sense by affirmation. Indeed its sense is just what is affirmed. And the same applies to negation, etc.

4.0641 One could say that negation must be related to the logical place determined by the negated proposition.

The negating proposition determines a logical place *different* from that of the negated proposition.

The negating proposition determines a logical place with the help of the logical place of the negated proposition. For it describes it as lying outside the latter's logical place.

The negated proposition can be negated again, and this in itself shows that what is negated is already a proposition, and not merely something that is preliminary to a proposition.

4.1 Propositions represent the existence and non-existence of states of affairs.

4.11 Die Gesamtheit der wahren Sätze ist die gesamte Naturwissenschaft (oder die Gesamtheit der Naturwissenschaften).

4.111 Die Philosophie ist keine der Naturwissenschaften.

(Das Wort „Philosophie" muß etwas bedeuten, was über oder unter, aber nicht neben den Naturwissenschaften steht.)

4.112 Der Zweck der Philosophie ist die logische Klärung der Gedanken.

Die Philosophie ist keine Lehre, sondern eine Tätigkeit.

Ein philosophisches Werk besteht wesentlich aus Erläuterungen.

Das Resultat der Philosophie sind nicht „philosophische Sätze", sondern das Klarwerden von Sätzen.

Die Philosophie soll die Gedanken, die sonst, gleichsam, trübe und verschwommen sind, klar machen und scharf abgrenzen.

4.1121 Die Psychologie ist der Philosophie nicht verwandter als irgend eine andere Naturwissenschaft.

Erkenntnistheorie ist die Philosophie der Psychologie.

Entspricht nicht mein Studium der Zeichensprache dem Studium der Denkprozesse, welches die Philosophen für die Philosophie der Logik für so wesentlich hielten? Nur verwickelten sie sich meistens in unwesentliche psychologische Untersuchungen und eine analoge Gefahr gibt es auch bei meiner Methode.

4.1122 Die Darwinsche Theorie hat mit der Philosophie nicht mehr zu schaffen als irgend eine andere Hypothese der Naturwissenschaft.

4.113 Die Philosophie begrenzt das bestreitbare Gebiet der Naturwissenschaft.

4.114 Sie soll das Denkbare abgrenzen und damit das Undenkbare.

Sie soll das Undenkbare von innen durch das Denkbare begrenzen.

4.115 Sie wird das Unsagbare bedeuten, indem sie das Sagbare klar darstellt.

4.11 The totality of true propositions is the whole of natural science (or the whole corpus of the natural sciences).

4.111 Philosophy is not one of the natural sciences.

(The word 'philosophy' must mean something whose place is above or below the natural sciences, not beside them.)

4.112 Philosophy aims at the logical clarification of thoughts.

Philosophy is not a body of doctrine but an activity.

A philosophical work consists essentially of elucidations.

Philosophy does not result in 'philosophical propositions', but rather in the clarification of propositions.

Without philosophy thoughts are, as it were, cloudy and indistinct: its task is to make them clear and to give them sharp boundaries.

4.1121 Psychology is no more closely related to philosophy than any other natural science.

Theory of knowledge is the philosophy of psychology.

Does not my study of sign-language correspond to the study of thought-processes, which philosophers used to consider so essential to the philosophy of logic? Only in most cases they got entangled in unessential psychological investigations, and with my method too there is an analogous risk.

4.1122 Darwin's theory has no more to do with philosophy than any other hypothesis in natural science.

4.113 Philosophy sets limits to the much disputed sphere of natural science.

4.114 It must set limits to what can be thought; and, in doing so, to what cannot be thought.

It must set limits to what cannot be thought by working outwards through what can be thought.

4.115 It will signify what cannot be said, by presenting clearly what can be said.

4.116 Alles, was überhaupt gedacht werden kann, kann klar gedacht werden. Alles, was sich aussprechen läßt, läßt sich klar aussprechen.

4.12 Der Satz kann die gesamte Wirklichkeit darstellen, aber er kann nicht das darstellen, was er mit der Wirklichkeit gemein haben muß, um sie darstellen zu können,— die logische Form.

Um die logische Form darstellen zu können, müßten wir uns mit dem Satze außerhalb der Logik aufstellen können, das heißt außerhalb der Welt.

4.121 Der Satz kann die logische Form nicht darstellen, sie spiegelt sich in ihm.

Was sich in der Sprache spiegelt, kann sie nicht darstellen.

Was *sich* in der Sprache ausdrückt, können *wir* nicht durch sie ausdrücken.

Der Satz *zeigt* die logische Form der Wirklichkeit.

Er weist sie auf.

4.1211 So zeigt ein Satz „fa", daß in seinem Sinn der Gegenstand a vorkommt, zwei Sätze „fa" und „ga", daß in ihnen beiden von demselben Gegenstand die Rede ist.

Wenn zwei Sätze einander widersprechen, so zeigt dies ihre Struktur; ebenso, wenn einer aus dem anderen folgt. U.s.w.

4.1212 Was gezeigt werden *kann*, *kann* nicht gesagt werden.

4.1213 Jetzt verstehen wir auch unser Gefühl: daß wir im Besitze einer richtigen logischen Auffassung seien, wenn nur einmal alles in unserer Zeichensprache stimmt.

4.122 Wir können in gewissem Sinne von formalen Eigenschaften der Gegenstände und Sachverhalte bezw. von Eigenschaften der Struktur der Tatsachen reden, und in demselben Sinne von formalen Relationen und Relationen von Strukturen.

(Statt Eigenschaft der Struktur sage ich auch „interne Eigenschaft"; statt Relation der Strukturen „interne Relation".

Ich führe diese Ausdrücke ein, um den Grund der bei den Philosophen sehr verbreiteten Verwechslung zwischen

4.116 Everything that can be thought at all can be thought clearly. Everything that can be put into words can be put clearly.

4.12 Propositions can represent the whole of reality, but they cannot represent what they must have in common with reality in order to be able to represent it—logical form.

In order to be able to represent logical form, we should have to be able to station ourselves with propositions somewhere outside logic, that is to say outside the world.

4.121 Propositions cannot represent logical form: it is mirrored in them.

What finds its reflection in language, language cannot represent.

What expresses *itself* in language, *we* cannot express by means of language.

Propositions *show* the logical form of reality.

They display it.

4.1211 Thus one proposition '*fa*' shows that the object *a* occurs in its sense, two propositions '*fa*' and '*ga*' show that the same object is mentioned in both of them.

If two propositions contradict one another, then their structure shows it; the same is true if one of them follows from the other. And so on.

4.1212 What *can* be shown, *cannot* be said.

4.1213 Now, too, we understand our feeling that once we have a sign-language in which everything is all right, we already have a correct logical point of view.

4.122 In a certain sense we can talk about formal properties of objects and states of affairs, or, in the case of facts, about structural properties: and in the same sense about formal relations and structural relations.

(Instead of 'structural property' I also say 'internal property'; instead of 'structural relation', 'internal relation'.

I introduce these expressions in order to indicate the source of the confusion between internal relations and relations proper (external relations), which is very widespread among philosophers.)

den internen Relationen und den eigentlichen (externen) Relationen zu zeigen.)

Das Bestehen solcher interner Eigenschaften und Relationen kann aber nicht durch Sätze behauptet werden, sondern es zeigt sich in den Sätzen, welche jene Sachverhalte darstellen und von jenen Gegenständen handeln.

4.1221 Eine interne Eigenschaft einer Tatsache können wir auch einen Zug dieser Tatsache nennen. (In dem Sinn, in welchem wir etwa von Gesichtszügen sprechen.)

4.123 Eine Eigenschaft ist intern, wenn es undenkbar ist, daß ihr Gegenstand sie nicht besitzt.

(Diese blaue Farbe und jene stehen in der internen Relation von heller und dunkler eo ipso. Es ist undenkbar, daß d i e s e beiden Gegenstände nicht in dieser Relation stünden.)

(Hier entspricht dem schwankenden Gebrauch der Worte „Eigenschaft" und „Relation" der schwankende Gebrauch des Wortes „Gegenstand".)

4.124 Das Bestehen einer internen Eigenschaft einer möglichen Sachlage wird nicht durch einen Satz ausgedrückt, sondern es drückt sich in dem sie darstellenden Satz durch eine interne Eigenschaft dieses Satzes aus.

Es wäre ebenso unsinnig, dem Satze eine formale Eigenschaft zuzusprechen, als sie ihm abzusprechen.

4.1241 Formen kann man nicht dadurch von einander unterscheiden, daß man sagt, die eine habe diese, die andere aber jene Eigenschaft; denn dies setzt voraus, daß es einen Sinn habe, beide Eigenschaften von beiden Formen auszusagen.

4.125 Das Bestehen einer internen Relation zwischen möglichen Sachlagen drückt sich sprachlich durch eine interne Relation zwischen den sie darstellenden Sätzen aus.

4.1251 Hier erledigt sich nun die Streitfrage, „ob alle Relationen intern oder extern seien".

4.1252 Reihen, welche durch i n t e r n e Relationen geordnet sind, nenne ich Formenreihen.

It is impossible, however, to assert by means of propositions that such internal properties and relations obtain: rather, this makes itself manifest in the propositions that represent the relevant states of affairs and are concerned with the relevant objects.

4.1221 An internal property of a fact can also be called a feature of that fact (in the sense in which we speak of facial features, for example).

4.123 A property is internal if it is unthinkable that its object should not possess it.

(This shade of blue and that one stand, eo ipso, in the internal relation of lighter to darker. It is unthinkable that *these* two objects should not stand in this relation.)

(Here the shifting use of the word 'object' corresponds to the shifting use of the words 'property' and 'relation'.)

4.124 The existence of an internal property of a possible situation is not expressed by means of a proposition: rather, it expresses itself in the proposition representing the situation, by means of an internal property of that proposition.

It would be just as nonsensical to assert that a proposition had a formal property as to deny it.

4.1241 It is impossible to distinguish forms from one another by saying that one has this property and another that property: for this presupposes that it makes sense to ascribe either property to either form.

4.125 The existence of an internal relation between possible situations expresses itself in language by means of an internal relation between the propositions representing them.

4.1251 Here we have the answer to the vexed question 'whether all relations are internal or external'.

4.1252 I call a series that is ordered by an *internal* relation a series of forms.

Die Zahlenreihe ist nicht nach einer externen, sondern nach einer internen Relation geordnet.

Ebenso die Reihe der Sätze

„aRb“,
„$(\exists x)$:aRx.xRb“,
„$(\exists x,y)$:aRx.xRy.yRb“,
u.s.f.

(Steht b in einer dieser Beziehungen zu a, so nenne ich b einen Nachfolger von a.)

4.126 In dem Sinne, in welchem wir von formalen Eigenschaften sprechen, können wir nun auch von formalen Begriffen reden.

(Ich führe diesen Ausdruck ein, um den Grund der Verwechslung der formalen Begriffe mit den eigentlichen Begriffen, welche die ganze alte Logik durchzieht, klar zu machen.)

Daß etwas unter einen formalen Begriff als dessen Gegenstand fällt, kann nicht durch einen Satz ausgedrückt werden. Sondern es zeigt sich an dem Zeichen dieses Gegenstandes selbst. (Der Name zeigt, daß er einen Gegenstand bezeichnet, das Zahlenzeichen, daß es eine Zahl bezeichnet, etc.)

Die formalen Begriffe können ja nicht, wie die eigentlichen Begriffe, durch eine Funktion dargestellt werden.

Denn ihre Merkmale, die formalen Eigenschaften, werden nicht durch Funktionen ausgedrückt.

Der Ausdruck der formalen Eigenschaft ist ein Zug gewisser Symbole.

Das Zeichen der Merkmale eines formalen Begriffes ist also ein charakteristischer Zug aller Symbole, deren Bedeutungen unter den Begriff fallen.

Der Ausdruck des formalen Begriffes, also, eine Satzvariable, in welcher nur dieser charakteristische Zug konstant ist.

4.127 Die Satzvariable bezeichnet den formalen Begriff und ihre Werte die Gegenstände, welche unter diesen Begriff fallen.

The order of the number-series is not governed by an external relation but by an internal relation.

The same is true of the series of propositions

$$
\begin{gathered}
`aRb\text{'}, \\
`(\exists x):aRx.xRb\text{'}, \\
`(\exists x,y):aRx.xRy.yRb\text{'}, \\
\text{and so forth.}
\end{gathered}
$$

(If *b* stands in one of these relations to *a*, I call *b* a successor of *a*.)

4.126 We can now talk about formal concepts, in the same sense that we speak of formal properties.

(I introduce this expression in order to exhibit the source of the confusion between formal concepts and concepts proper, which pervades the whole of traditional logic.)

When something falls under a formal concept as one of its objects, this cannot be expressed by means of a proposition. Instead it is shown in the very sign for this object. (A name shows that it signifies an object, a sign for a number that it signifies a number, etc.)

Formal concepts cannot, in fact, be represented by means of a function, as concepts proper can.

For their characteristics, formal properties, are not expressed by means of functions.

The expression for a formal property is a feature of certain symbols.

So the sign for the characteristics of a formal concept is a distinctive feature of all symbols whose meanings fall under the concept.

So the expression for a formal concept is a propositional variable in which this distinctive feature alone is constant.

4.127 The propositional variable signifies the formal concept, and its values signify the objects that fall under the concept.

4.1271 Jede Variable ist das Zeichen eines formalen Begriffes.

Denn jede Variable stellt eine konstante Form dar, welche alle ihre Werte besitzen, und die als formale Eigenschaft dieser Werte aufgefaßt werden kann.

4.1272 So ist der variable Name „x“ das eigentliche Zeichen des Scheinbegriffes *Gegenstand*.

Wo immer das Wort „Gegenstand“ („Ding“, „Sache“, etc.) richtig gebraucht wird, wird es in der Begriffsschrift durch den variablen Namen ausgedrückt.

Zum Beispiel in dem Satz: „Es gibt 2 Gegenstände, welche . . .“, durch „$(\exists x,y)$ ...“.

Wo immer es anders, also als eigentliches Begriffswort, gebraucht wird, entstehen unsinnige Scheinsätze.

So kann man z. B. nicht sagen: „Es gibt Gegenstände“, wie man etwa sagt: „Es gibt Bücher“. Und ebenso wenig: „Es gibt 100 Gegenstände“, oder: „Es gibt $\aleph_0$ Gegenstände“.

Und es ist unsinnig, von der *Anzahl aller Gegenstände* zu sprechen.

Dasselbe gilt von den Worten „Komplex“, „Tatsache“, „Funktion“, „Zahl“, etc.

Sie alle bezeichnen formale Begriffe und werden in der Begriffsschrift durch Variable, nicht durch Funktionen oder Klassen dargestellt. (Wie Frege und Russell glaubten.)

Ausdrücke wie: „1 ist eine Zahl“, „Es gibt nur Eine Null“, und alle ähnlichen sind unsinnig.

(Es ist ebenso unsinnig zu sagen: „Es gibt nur Eine 1“, als es unsinnig wäre, zu sagen: „2+2 ist um 3 Uhr gleich 4“.)

4.12721 Der formale Begriff ist mit einem Gegenstand, der unter ihn fällt, bereits gegeben. Man kann also nicht Gegenstände eines formalen Begriffes *und* den formalen Begriff selbst als Grundbegriffe einführen. Man kann also z. B. nicht den Begriff der Funktion, und auch spezielle Funktionen (wie Russell) als Grundbegriffe einführen; oder den Begriff der Zahl und bestimmte Zahlen.

4.1273 Wollen wir den allgemeinen Satz: „b ist ein Nachfolger von a“, in der Begriffsschrift ausdrücken, so brauchen

4.1271 Every variable is the sign for a formal concept.

For every variable represents a constant form that all its values possess, and this can be regarded as a formal property of those values.

4.1272 Thus the variable name 'x' is the proper sign for the pseudo-concept *object*.

Wherever the word 'object' ('thing', etc.) is correctly used, it is expressed in conceptual notation by a variable name.

For example, in the proposition, 'There are 2 objects which . . .', it is expressed by '$(\exists x, y) \ldots$'.

Wherever it is used in a different way, that is as a proper concept-word, nonsensical pseudo-propositions are the result.

So one cannot say, for example, 'There are objects', as one might say, 'There are books'. And it is just as impossible to say, 'There are 100 objects', or, 'There are $\aleph_0$ objects'.

And it is nonsensical to speak of the *total number of objects*.

The same applies to the words 'complex', 'fact', 'function', 'number', etc.

They all signify formal concepts, and are represented in conceptual notation by variables, not by functions or classes (as Frege and Russell believed).

'1 is a number', 'There is only one zero', and all similar expressions are nonsensical.

(It is just as nonsensical to say, 'There is only one 1', as it would be to say, '2+2 at 3 o'clock equals 4'.)

4.12721 A formal concept is given immediately any object falling under it is given. It is not possible, therefore, to introduce as primitive ideas objects belonging to a formal concept *and* the formal concept itself. So it is impossible, for example, to introduce as primitive ideas both the concept of a function and specific functions, as Russell does; or the concept of a number and particular numbers.

4.1273 If we want to express in conceptual notation the general proposition, 'b is a successor of a', then we require

wir hierzu einen Ausdruck für das allgemeine Glied der Formenreihe:

$$aRb,$$
$$(\exists x):aRx.xRb,$$
$$(\exists x,y):aRx.xRy.yRb,$$
$$\ldots\,.$$

Das allgemeine Glied einer Formenreihe kann man nur durch eine Variable ausdrücken, denn der Begriff: Glied dieser Formenreihe, ist ein formaler Begriff. (Dies haben Frege und Russell übersehen; die Art und Weise, wie sie allgemeine Sätze wie den obigen ausdrücken wollen, ist daher falsch; sie enthält einen circulus vitiosus.)

Wir können das allgemeine Glied der Formenreihe bestimmen, indem wir ihr erstes Glied angeben und die allgemeine Form der Operation, welche das folgende Glied aus dem vorhergehenden Satz erzeugt.

4.1274 Die Frage nach der Existenz eines formalen Begriffes ist unsinnig. Denn kein Satz kann eine solche Frage beantworten.

(Man kann also z. B. nicht fragen: „Gibt es unanalysierbare Subjekt-Prädikatsätze?")

4.128 Die logischen Formen sind zahllos.

Darum gibt es in der Logik keine ausgezeichneten Zahlen und darum gibt es keinen philosophischen Monismus oder Dualismus, etc.

4.2 Der Sinn des Satzes ist seine Übereinstimmung und Nichtübereinstimmung mit den Möglichkeiten des Bestehens und Nichtbestehens der Sachverhalte.

4.21 Der einfachste Satz, der Elementarsatz, behauptet das Bestehen eines Sachverhaltes.

4.211 Ein Zeichen des Elementarsatzes ist es, daß kein Elementarsatz mit ihm in Widerspruch stehen kann.

4.22 Der Elementarsatz besteht aus Namen. Er ist ein Zusammenhang, eine Verkettung, von Namen.

4.221 Es ist offenbar, daß wir bei der Analyse der Sätze auf Elementarsätze kommen müssen, die aus Namen in unmittelbarer Verbindung bestehen.

an expression for the general term of the series of forms

$$aRb,$$
$$(\exists x):aRx.xRb,$$
$$(\exists x,y):aRx.xRy.yRb,$$
$$\ldots.$$

In order to express the general term of a series of forms, we must use a variable, because the concept 'term of that series of forms' is a *formal* concept. (This is what Frege and Russell overlooked: consequently the way in which they want to express general propositions like the one above is incorrect; it contains a vicious circle.)

We can determine the general term of a series of forms by giving its first term and the general form of the operation that produces the next term out of the proposition that precedes it.

4.1274 To ask whether a formal concept exists is nonsensical. For no proposition can be the answer to such a question.

(So, for example, the question, 'Are there unanalysable subject-predicate propositions?' cannot be asked.)

4.128 Logical forms are *without* number.

Hence there are no pre-eminent numbers in logic, and hence there is no possibility of philosophical monism or dualism, etc.

4.2 The sense of a proposition is its agreement and disagreement with possibilities of existence and non-existence of states of affairs.

4.21 The simplest kind of proposition, an elementary proposition, asserts the existence of a state of affairs.

4.211 It is a sign of a proposition's being elementary that there can be no elementary proposition contradicting it.

4.22 An elementary proposition consists of names. It is a nexus, a concatenation, of names.

4.221 It is obvious that the analysis of propositions must bring us to elementary propositions which consist of names in immediate combination.

Es frägt sich hier, wie kommt der Satzverband zustande.

4.2211 Auch wenn die Welt unendlich komplex ist, so daß jede Tatsache aus unendlich vielen Sachverhalten besteht und jeder Sachverhalt aus unendlich vielen Gegenständen zusammengesetzt ist, auch dann müßte es Gegenstände und Sachverhalte geben.

4.23 Der Name kommt im Satz nur im Zusammenhange des Elementarsatzes vor.

4.24 Die Namen sind die einfachen Symbole, ich deute sie durch einzelne Buchstaben („x", „y", „z") an.

Den Elementarsatz schreibe ich als Funktion der Namen in der Form: „fx", „ϕ(x,y)", etc.

Oder ich deute ihn durch die Buchstaben p, q, r an.

4.241 Gebrauche ich zwei Zeichen in ein und derselben Bedeutung, so drücke ich dies aus, indem ich zwischen beide das Zeichen „=" setze.

„a = b" heißt also: Das Zeichen „a" ist durch das Zeichen „b" ersetzbar.

(Führe ich durch eine Gleichung ein neues Zeichen „b" ein, indem ich bestimme, es solle ein bereits bekanntes Zeichen „a" ersetzen, so schreibe ich die Gleichung — Definition — (wie Russell) in der Form „a = b Def." Die Definition ist eine Zeichenregel.)

4.242 Ausdrücke von der Form „a = b" sind also nur Behelfe der Darstellung; sie sagen nichts über die Bedeutung der Zeichen „a", „b" aus.

4.243 Können wir zwei Namen verstehen, ohne zu wissen, ob sie dasselbe Ding oder zwei verschiedene Dinge bezeichnen? — Können wir einen Satz, worin zwei Namen vorkommen, verstehen, ohne zu wissen, ob sie Dasselbe oder Verschiedenes bedeuten?

Kenne ich etwa die Bedeutung eines englischen und eines gleichbedeutenden deutschen Wortes, so ist es unmöglich, daß ich nicht weiß, daß die beiden gleichbedeutend sind; es ist unmöglich, daß ich sie nicht ineinander übersetzen kann.

Ausdrücke wie „a = a", oder von diesen abgeleitete,

This raises the question how such combination into propositions comes about.

4.2211 Even if the world is infinitely complex, so that every fact consists of infinitely many states of affairs and every state of affairs is composed of infinitely many objects, there would still have to be objects and states of affairs.

4.23 It is only in the nexus of an elementary proposition that a name occurs in a proposition.

4.24 Names are the simple symbols: I indicate them by single letters (‘*x*’, ‘*y*’, ‘*z*’).

I write elementary propositions as functions of names, so that they have the form ‘*fx*’, ‘$\phi(x,y)$’, etc.

Or I indicate them by the letters ‘*p*’, ‘*q*’, ‘*r*’.

4.241 When I use two signs with one and the same meaning, I express this by putting the sign ‘=’ between them.

So ‘$a = b$’ means that the sign ‘*b*’ can be substituted for the sign ‘*a*’.

(If I use an equation to introduce a new sign ‘*b*’, laying down that it shall serve as a substitute for a sign ‘*a*’ that is already known, then, like Russell, I write the equation—definition—in the form ‘$a = b$ Def.’ A definition is a rule dealing with signs.)

4.242 Expressions of the form ‘$a = b$’ are, therefore, mere representational devices. They state nothing about the meaning of the signs ‘*a*’ and ‘*b*’.

4.243 Can we understand two names without knowing whether they signify the same thing or two different things?—Can we understand a proposition in which two names occur without knowing whether their meaning is the same or different?

Suppose I know the meaning of an English word and of a German word that means the same: then it is impossible for me to be unaware that they do mean the same; I must be capable of translating each into the other.

Expressions like ‘$a = a$’, and those derived from them, are neither elementary propositions nor is there any other

sind weder Elementarsätze, noch sonst sinnvolle Zeichen. (Dies wird sich später zeigen.)

4.25 Ist der Elementarsatz wahr, so besteht der Sachverhalt; ist der Elementarsatz falsch, so besteht der Sachverhalt nicht.

4.26 Die Angabe aller wahren Elementarsätze beschreibt die Welt vollständig. Die Welt ist vollständig beschrieben durch die Angaben aller Elementarsätze plus der Angabe, welche von ihnen wahr und welche falsch sind.

4.27 Bezüglich des Bestehens und Nichtbestehens von n Sachverhalten gibt es $K_n = \sum_{\nu=0}^{n} \binom{n}{\nu}$ Möglichkeiten.

Es können alle Kombinationen der Sachverhalte bestehen, die andern nicht bestehen.

4.28 Diesen Kombinationen entsprechen ebenso viele Möglichkeiten der Wahrheit — und Falschheit — von n Elementarsätzen.

4.3 Die Wahrheitsmöglichkeiten der Elementarsätze bedeuten die Möglichkeiten des Bestehens und Nichtbestehens der Sachverhalte.

4.31 Die Wahrheitsmöglichkeiten können wir durch Schemata folgender Art darstellen („W" bedeutet „wahr", „F" „falsch"; die Reihen der „W" und „F" unter der Reihe der Elementarsätze bedeuten in leichtverständlicher Symbolik deren Wahrheitsmöglichkeiten):

p	q	r
W	W	W
F	W	W
W	F	W
W	W	F
F	F	W
F	W	F
W	F	F
F	F	F

,

p	q
W	W
F	W
W	F
F	F

,

p
W
F

.

4.4 Der Satz ist der Ausdruck der Übereinstimmung und Nichtübereinstimmung mit den Wahrheitsmöglichkeiten der Elementarsätze.

way in which they have sense. (This will become evident later.)

4.25 If an elementary proposition is true, the state of affairs exists: if an elementary proposition is false, the state of affairs does not exist.

4.26 If all true elementary propositions are given, the result is a complete description of the world. The world is completely described by giving all elementary propositions, and adding which of them are true and which false.

4.27 For n states of affairs, there are $K_n = \sum_{\nu=0}^{n} \binom{n}{\nu}$ possibilities of existence and non-existence.

Of these states of affairs any combination can exist and the remainder not exist.

4.28 There correspond to these combinations the same number of possibilities of truth—and falsity—for n elementary propositions.

4.3 Truth-possibilities of elementary propositions mean possibilities of existence and non-existence of states of affairs.

4.31 We can represent truth-possibilities by schemata of the following kind ('*T*' means 'true', '*F*' means 'false'; the rows of '*T*'s' and '*F*'s' under the row of elementary propositions symbolize their truth-possibilities in a way that can easily be understood):

p	q	r
T	*T*	*T*
F	*T*	*T*
T	*F*	*T*
T	*T*	*F*
F	*F*	*T*
F	*T*	*F*
T	*F*	*F*
F	*F*	*F*

,

p	q
T	*T*
F	*T*
T	*F*
F	*F*

,

p
T
F

.

4.4 A proposition is an expression of agreement and disagreement with truth-possibilities of elementary propositions.

4.41 Die Wahrheitsmöglichkeiten der Elementarsätze sind die Bedingungen der Wahrheit und Falschheit der Sätze.

4.411 Es ist von vornherein wahrscheinlich, daß die Einführung der Elementarsätze für das Verständnis aller anderen Satzarten grundlegend ist. Ja, das Verständnis der allgemeinen Sätze hängt f ü h l b a r von dem der Elementarsätze ab.

4.42 Bezüglich der Übereinstimmung und Nichtübereinstimmung eines Satzes mit den Wahrheitsmöglichkeiten von n Elementarsätzen gibt es $\sum_{\kappa=0}^{K_n} \binom{K_n}{\kappa} = L_n$ Moglichkeiten.

4.43 Die Übereinstimmung mit den Wahrheitsmöglichkeiten können wir dadurch ausdrücken, indem wir ihnen im Schema etwa das Abzeichen „W" (wahr) zuordnen.

Das Fehlen dieses Abzeichens bedeutet die Nichtübereinstimmung.

4.431 Der Ausdruck der Übereinstimmung und Nichtübereinstimmung mit den Wahrheitsmöglichkeiten der Elementarsätze drückt die Wahrheitsbedingungen des Satzes aus.

Der Satz ist der Ausdruck seiner Wahrheitsbedingungen.

(Frege hat sie daher ganz richtig als Erklärung der Zeichen seiner Begriffsschrift vorausgeschickt..Nur ist die Erklärung des Wahrheitsbegriffes bei Frege falsch: Wären „das Wahre" und „das Falsche" wirklich Gegenstände und die Argumente in ~p etc., dann wäre nach Freges Bestimmung der Sinn von „~p" keineswegs bestimmt.)

4.44 Das Zeichen, welches durch die Zuordnung jener Abzeichen „W" und der Wahrheitsmöglichkeiten entsteht, ist ein Satzzeichen.

4.441 Es ist klar, daß dem Komplex der Zeichen „F" und „W" kein Gegenstand (oder Komplex von Gegenständen) entspricht; so wenig, wie den horizontalen und vertikalen Strichen oder den Klammern.—„Logische Gegenstände" gibt es nicht.

Analoges gilt natürlich für alle Zeichen, die dasselbe ausdrücken wie die Schemata der „W" und „F".

4.41 Truth-possibilities of elementary propositions are the conditions of the truth and falsity of propositions.

4.411 It immediately strikes one as probable that the introduction of elementary propositions provides the basis for understanding all other kinds of proposition. Indeed the understanding of general propositions *palpably* depends on the understanding of elementary propositions.

4.42 For n elementary propositions there are $\sum_{\kappa=0}^{K_n} \binom{K_n}{\kappa} = L_n$ ways in which a proposition can agree and disagree with their truth-possibilities

4.43 We can express agreement with truth-possibilities by correlating the mark '*T*' (true) with them in the schema.

The absence of this mark means disagreement.

4.431 The expression of agreement and disagreement with the truth-possibilities of elementary propositions expresses the truth-conditions of a proposition.

A proposition is the expression of its truth-conditions.

(Thus Frege was quite right to use them as a starting point when he explained the signs of his conceptual notation. But the explanation of the concept of truth that Frege gives is mistaken: if 'the true' and 'the false' were really objects, and were the arguments in $\sim p$ etc., then Frege's method of determining the sense of '$\sim p$' would leave it absolutely undetermined.)

4.44 The sign that results from correlating the mark '*T*' with truth-possibilities is a propositional sign.

4.441 It is clear that a complex of the signs '*F*' and '*T*' has no object (or complex of objects) corresponding to it, just as there is none corresponding to the horizontal and vertical lines or to the brackets.—There are no 'logical objects'.

Of course the same applies to all signs that express what the schemata of '*T*'s' and '*F*'s' express.

4.442 Es ist z. B.

„p	q	“
W	W	W
F	W	W
W	F	
„F	F	W

ein Satzzeichen.

(Freges „Urteilstrich“ „|–“ ist logisch ganz bedeutungslos; er zeigt bei Frege (und Russell) nur an, daß diese Autoren die so bezeichneten Sätze für wahr halten. „|–“ gehört daher ebensowenig zum Satzgefüge wie etwa die Nummer des Satzes. Ein Satz kann unmöglich von sich selbst aussagen, daß er wahr ist.)

Ist die Reihenfolge der Wahrheitsmöglichkeiten im Schema durch eine Kombinationsregel ein für allemal festgesetzt, dann ist die letzte Kolonne allein schon ein Ausdruck der Wahrheitsbedingungen. Schreiben wir diese Kolonne als Reihe hin, so wird das Satzzeichen zu

„(WW-W) (p,q)“,

oder deutlicher

„(WWFW) (p,q)“.

(Die Anzahl der Stellen in der linken Klammer ist durch die Anzahl der Glieder in der rechten bestimmt.)

4.45 Für n Elementarsätze gibt es L_n mögliche Gruppen von Wahrheitsbedingungen.

Die Gruppen von Wahrheitsbedingungen, welche zu den Wahrheitsmöglichkeiten einer Anzahl von Elementarsätzen gehören, lassen sich in eine Reihe ordnen.

4.46 Unter den möglichen Gruppen von Wahrheitsbedingungen gibt es zwei extreme Fälle.

In dem einen Fall ist der Satz für sämtliche Wahrheitsmöglichkeiten der Elementarsätze wahr. Wir sagen, die Wahrheitsbedingungen sind t a u t o l o g i s c h.

4.442 For example, the following is a propositional sign:

‘p	q	’
T	T	T
F	T	T
T	F	
F	F	T.

(Frege's ‘judgement-stroke’ ‘|–’ is logically quite meaningless: in the works of Frege (and Russell) it simply indicates that these authors hold the propositions marked with this sign to be true. Thus ‘|–’ is no more a component part of a proposition than is, for instance, the proposition's number. It is quite impossible for a proposition to state that it itself is true.)

If the order of the truth-possibilities in a schema is fixed once and for all by a combinatory rule, then the last column by itself will be an expression of the truth-conditions. If we now write this column as a row, the propositional sign will become

‘$(TT\text{-}T)\,(p,q)$’

or more explicitly

‘$(TTFT)\,(p,q)$’.

(The number of places in the left-hand pair of brackets is determined by the number of terms in the right-hand pair.)

4.45 For n elementary propositions there are L_n possible groups of truth-conditions.

The groups of truth-conditions that are obtainable from the truth-possibilities of a given number of elementary propositions can be arranged in a series.

4.46 Among the possible groups of truth-conditions there are two extreme cases.

In one of these cases the proposition is true for all the truth-possibilities of the elementary propositions. We say that the truth-conditions are *tautological*.

Im zweiten Fall ist der Satz für sämtliche Wahrheitsmöglichkeiten falsch: Die Wahrheitsbedingungen sind kontradiktorisch.

Im ersten Fall nennen wir den Satz eine Tautologie, im zweiten Fall eine Kontradiktion.

4.461 Der Satz zeigt, was er sagt, die Tautologie und die Kontradiktion, daß sie nichts sagen.

Die Tautologie hat keine Wahrheitsbedingungen, denn sie ist bedingungslos wahr; und die Kontradiktion ist unter keiner Bedingung wahr.

Tautologie und Kontradiktion sind sinnlos.

(Wie der Punkt, von dem zwei Pfeile in entgegengesetzter Richtung auseinandergehen.)

(Ich weiß z. B. nichts über das Wetter, wenn ich weiß, daß es regnet oder nicht regnet.)

4.4611 Tautologie und Kontradiktion sind aber nicht unsinnig; sie gehören zum Symbolismus, und zwar ähnlich wie die „0" zum Symbolismus der Arithmetik.

4.462 Tautologie und Kontradiktion sind nicht Bilder der Wirklichkeit. Sie stellen keine mögliche Sachlage dar. Denn jene läßt jede mögliche Sachlage zu, diese keine.

In der Tautologie heben die Bedingungen der Übereinstimmung mit der Welt — die darstellenden Beziehungen — einander auf, so daß sie in keiner darstellenden Beziehung zur Wirklichkeit steht.

4.463 Die Wahrsheitsbedingungen bestimmen den Spielraum, der den Tatsachen durch den Satz gelassen wird.

(Der Satz, das Bild, das Modell, sind im negativen Sinne wie ein fester Körper, der die Bewegungsfreiheit der anderen beschränkt; im positiven Sinne, wie der von fester Substanz begrenzte Raum, worin ein Körper Platz hat.)

Die Tautologie läßt der Wirklichkeit den ganzen — unendlichen — logischen Raum; die Kontradiktion erfüllt den ganzen logischen Raum und läßt der Wirklichkeit keinen Punkt. Keine von beiden kann daher die Wirklichkeit irgendwie bestimmen.

4.464 Die Wahrheit der Tautologie ist gewiß, des Satzes möglich, der Kontradiktion unmöglich.

(Gewiß, möglich, unmöglich: Hier haben wir das

In the second case the proposition is false for all the truth-possibilities: the truth-conditions are *contradictory*.

In the first case we call the proposition a tautology; in the second, a contradiction.

4.461 Propositions show what they say: tautologies and contradictions show that they say nothing.

A tautology has no truth-conditions, since it is unconditionally true: and a contradiction is true on no condition.

Tautologies and contradictions lack sense.

(Like a point from which two arrows go out in opposite directions to one another.)

(For example, I know nothing about the weather when I know that it is either raining or not raining.)

4.4611 Tautologies and contradictions are not, however, nonsensical. They are part of the symbolism, much as '0' is part of the symbolism of arithmetic.

4.462 Tautologies and contradictions are not pictures of reality. They do not represent any possible situations. For the former admit *all* possible situations, and the latter *none*.

In a tautology the conditions of agreement with the world—the representational relations—cancel one another, so that it does not stand in any representational relation to reality.

4.463 The truth-conditions of a proposition determine the range that it leaves open to the facts.

(A proposition, a picture, or a model is, in the negative sense, like a solid body that restricts the freedom of movement of others, and, in the positive sense, like a space bounded by solid substance in which there is room for a body.)

A tautology leaves open to reality the whole—the infinite whole—of logical space: a contradiction fills the whole of logical space leaving no point of it for reality. Thus neither of them can determine reality in any way.

4.464 A tautology's truth is certain, a proposition's possible, a contradiction's impossible.

(Certain, possible, impossible: here we have the first

Anzeichen jener Gradation, die wir in der Wahrscheinlichkeitslehre brauchen.)

4.465 Das logische Produkt einer Tautologie und eines Satzes sagt dasselbe, wie der Satz. Also ist jenes Produkt identisch mit dem Satz. Denn man kann das Wesentliche des Symbols nicht ändern, ohne seinen Sinn zu ändern.

4.466 Einer bestimmten logischen Verbindung von Zeichen entspricht eine bestimmte logische Verbindung ihrer Bedeutungen; *jede beliebige* Verbindung entspricht nur den unverbundenen Zeichen.

Das heißt, Sätze, die für jede Sachlage wahr sind, können überhaupt keine Zeichenverbindungen sein, denn sonst könnten ihnen nur bestimmte Verbindungen von Gegenständen entsprechen.

(Und keiner logischen Verbindung entspricht *keine* Verbindung der Gegenstände.)

Tautologie und Kontradiktion sind die Grenzfälle der Zeichenverbindung, nämlich ihre Auflösung.

4.4661 Freilich sind auch in der Tautologie und Kontradiktion die Zeichen noch mit einander verbunden, d. h. sie stehen in Beziehungen zu einander, aber diese Beziehungen sind bedeutungslos, dem *Symbol* unwesentlich.

4.5 Nun scheint es möglich zu sein, die allgemeinste Satzform anzugeben: das heißt, eine Beschreibung der Sätze *irgend einer* Zeichensprache zu geben, so daß jeder mögliche Sinn durch ein Symbol, auf welches die Beschreibung paßt, ausgedrückt werden kann, und daß jedes Symbol, worauf die Beschreibung paßt, einen Sinn ausdrücken kann, wenn die Bedeutungen der Namen entsprechend gewählt werden.

Es ist klar, daß bei der Beschreibung der allgemeinsten Satzform *nur* ihr Wesentliches beschrieben werden darf, — sonst wäre sie nämlich nicht die allgemeinste.

Daß es eine allgemeine Satzform gibt, wird dadurch bewiesen, daß es keinen Satz geben darf, dessen Form man nicht hätte voraussehen (d. h. konstruieren) können. Die allgemeine Form des Satzes ist: Es verhält sich so und so.

indication of the scale that we need in the theory of probability.)

4.465 The logical product of a tautology and a proposition says the same thing as the proposition. This product, therefore, is identical with the proposition. For it is impossible to alter what is essential to a symbol without altering its sense.

4.466 What corresponds to a determinate logical combination of signs is a determinate logical combination of their meanings. It is only to the uncombined signs that *absolutely any* combination corresponds.

In other words, propositions that are true for every situation cannot be combinations of signs at all, since, if they were, only determinate combinations of objects could correspond to them.

(And what is not a logical combination has *no* combination of objects corresponding to it.)

Tautology and contradiction are the limiting cases—indeed the disintegration—of the combination of signs.

4.4661 Admittedly the signs are still combined with one another even in tautologies and contradictions—i.e. they stand in certain relations to one another: but these relations have no meaning, they are not essential to the *symbol*.

4.5 It now seems possible to give the most general propositional form: that is, to give a description of the propositions of *any* sign-language *whatsoever* in such a way that every possible sense can be expressed by a symbol satisfying the description, and every symbol satisfying the description can express a sense, provided that the meanings of the names are suitably chosen.

It is clear that *only* what is essential to the most general propositional form may be included in its description—for otherwise it would not be the most general form.

The existence of a general propositional form is proved by the fact that there cannot be a proposition whose form could not have been foreseen (i.e. constructed). The general form of a proposition is: This is how things stand.

4.51 Angenommen, mir wären *alle* Elementarsätze gegeben: Dann läßt sich einfach fragen: Welche Sätze kann ich aus ihnen bilden? Und das sind *alle* Sätze und *so* sind sie begrenzt.

4.52 Die Sätze sind alles, was aus der Gesamtheit aller Elementarsätze folgt (natürlich auch daraus, daß es die *Gesamtheit aller* ist). (So könnte man in gewissem Sinne sagen, daß *alle* Sätze Verallgemeinerungen der Elementarsätze sind.)

4.53 Die allgemeine Satzform ist eine Variable.

5 Der Satz ist eine Wahrheitsfunktion der Elementarsätze.

(Der Elementarsatz ist eine Wahrheitsfunktion seiner selbst.)

5.01 Die Elementarsätze sind die Wahrheitsargumente des Satzes.

5.02 Es liegt nahe, die Argumente von Funktionen mit den Indices von Namen zu verwechseln. Ich erkenne nämlich sowohl am Argument wie am Index die Bedeutung des sie enthaltenden Zeichens.

In Russells „$+_c$" ist z. B. „$_c$" ein Index, der darauf hinweist, daß das ganze Zeichen das Additionszeichen für Kardinalzahlen ist. Aber diese Bezeichnung beruht auf willkürlicher Übereinkunft und man könnte statt „$+_c$" auch ein einfaches Zeichen wählen; in „~p" aber ist „p" kein Index, sondern ein Argument: der Sinn von „~p" *kann nicht* verstanden werden, ohne daß vorher der Sinn von „p" verstanden worden wäre. (Im Namen Julius Cäsar ist „Julius" ein Index. Der Index ist immer ein Teil einer Beschreibung des Gegenstandes, dessen Namen wir ihn anhängen. Z. B. *der* Cäsar aus dem Geschlechte der Julier.)

Die Verwechslung von Argument und Index liegt, wenn ich mich nicht irre, der Theorie Freges von der Bedeutung der Sätze und Funktionen zugrunde. Für Frege waren die Sätze der Logik Namen, und deren Argumente die Indices dieser Namen.

5.1 Die Wahrheitsfunktionen lassen sich in Reihen ordnen.

Das ist die Grundlage der Wahrscheinlichkeitslehre.

4.51 Suppose that I am given *all* elementary propositions: then I can simply ask what propositions I can construct out of them. And there I have *all* propositions, and *that* fixes their limits.

4.52 Propositions comprise all that follows from the totality of all elementary propositions (and, of course, from its being the *totality* of them *all*). (Thus, in a certain sense, it could be said that *all* propositions were generalizations of elementary propositions.)

4.53 The general propositional form is a variable.

5 A proposition is a truth-function of elementary propositions.

(An elementary proposition is a truth-function of itself.)

5.01 Elementary propositions are the truth-arguments of propositions.

5.02 The arguments of functions are readily confused with the affixes of names. For both arguments and affixes enable me to recognize the meaning of the signs containing them.

For example, when Russell writes '$+_c$', the 'c' is an affix which indicates that the sign as a whole is the addition-sign for cardinal numbers. But the use of this sign is the result of arbitrary convention and it would be quite possible to choose a simple sign instead of '$+_c$'; in '$\sim p$', however, 'p' is not an affix but an argument: the sense of '$\sim p$' *cannot* be understood unless the sense of 'p' has been understood already. (In the name Julius Caesar 'Julius' is an affix. An affix is always part of a description of the object to whose name we attach it: e.g. *the* Caesar of the Julian gens.)

If I am not mistaken, Frege's theory about the meaning of propositions and functions is based on the confusion between an argument and an affix. Frege regarded the propositions of logic as names, and their arguments as the affixes of those names.

5.1 Truth-functions can be arranged in series.

That is the foundation of the theory of probability.

5.101 Die Wahrheitsfunktionen jeder Anzahl von Elementarsätzen lassen sich in einem Schema folgender Art hinschreiben:

(W W W W)	(p, q)	Tautologie	(Wenn p, so p; und wenn q, so q.) $(p \supset p \,.\, q \supset q)$
(F W W W)	(p, q)	in Worten:	Nicht beides p und q. $(\sim(p \,.\, q))$
(W F W W)	(p, q)	„ „ :	Wenn q, so p. $(q \supset p)$
(W W F W)	(p, q)	„ „ :	Wenn p, so q. $(p \supset q)$
(W W W F)	(p, q)	„ „ :	p oder q. $(p \vee q)$
(F F W W)	(p, q)	„ „ :	Nicht q. $(\sim q)$
(F W F W)	(p, q)	„ „ :	Nicht p. $(\sim p)$
(F W W F)	(p, q)	„ „ :	p oder q, aber nicht beide. $(p \,.\, \sim q : \vee : q \,.\, \sim p)$
(W F F W)	(p, q)	„ „ :	Wenn p, so q; und wenn q, so p. $(p \equiv q)$
(W F W F)	(p, q)	„ „ :	p
(W W F F)	(p, q)	„ „ :	q
(F F F W)	(p, q)	„ „ :	Weder p, noch q. $(\sim p \,.\, \sim q$ oder $p \vert q)$
(F F W F)	(p, q)	„ „ :	p und nicht q. $(p \,.\, \sim q)$
(F W F F)	(p, q)	„ „ :	q und nicht p. $(q \,.\, \sim p)$
(W F F F)	(p, q)	„ „ :	q und p. $(q \,.\, p)$
(F F F F)	(p, q)	Kontradiktion	(p und nicht p; und q und nicht q.) $(p \,.\, \sim p \,.\, q \,.\, \sim q)$

Diejenigen Wahrheitsmöglichkeiten seiner Wahrheitsargumente, welche den Satz bewahrheiten, will ich seine *Wahrheitsgründe* nennen.

5.11 Sind die Wahrheitsgründe, die einer Anzahl von Sätzen gemeinsam sind, sämtlich auch Wahrheitsgründe eines bestimmten Satzes, so sagen wir, die Wahrheit dieses Satzes folge aus der Wahrheit jener Sätze.

5.12 Insbesondere folgt die Wahrheit eines Satzes „p“ aus der Wahrheit eines anderen „q“, wenn alle Wahrheitsgründe des zweiten Wahrheitsgründe des ersten sind.

5.121 Die Wahrheitsgründe des einen sind in denen des anderen enthalten; p folgt aus q.

5.122 Folgt p aus q, so ist der Sinn von „p“ im Sinne von „q“ enthalten.

5.123 Wenn ein Gott eine Welt erschafft, worin gewisse Sätze wahr sind, so schafft er damit auch schon eine Welt, in welcher alle ihre Folgesätze stimmen. Und ähnlich könnte er keine Welt schaffen, worin der Satz „p“ wahr ist, ohne seine sämtlichen Gegenstände zu schaffen.

5.124 Der Satz bejaht jeden Satz, der aus ihm folgt.

5.1241 „p.q“ ist einer der Sätze, welche „p“ bejahen, und zugleich einer der Sätze, welche „q“ bejahen.

5.101 The truth-functions of a given number of elementary propositions can always be set out in a schema of the following kind:

(*T T T T*) (p, q) Tautology (If *p* then *p*, and if *q* then *q*.) $(p \supset p \,.\, q \supset q)$
(*F T T T*) (p, q) In words: Not both *p* and *q*. $(\sim(p \,.\, q))$
(*T F T T*) (p, q) „ „ : If *q* then *p*. $(q \supset p)$
(*T T F T*) (p, q) „ „ : If *p* then *q*. $(p \supset q)$
(*T T T F*) (p, q) „ „ : *p* or *q*. $(p \vee q)$
(*F F T T*) (p, q) „ „ : Not *q*. $(\sim q)$
(*F T F T*) (p, q) „ „ : Not *p*. $(\sim p)$
(*F T T F*) (p, q) „ „ : *p* or *q*, but not both. $(p \,.\, \sim q : \vee : q \,.\, \sim p)$
(*T F F T*) (p, q) „ „ : If *p* then *q*, and if *q* then *p*. $(p \equiv q)$
(*T F T F*) (p, q) „ „ : *p*
(*T T F F*) (p, q) „ „ : *q*
(*F F F T*) (p, q) „ „ : Neither *p* nor *q*. $(\sim p \,.\, \sim q$ or $p|q)$
(*F F T F*) (p, q) „ „ : *p* and not *q*. $(p \,.\, \sim q)$
(*F T F F*) (p, q) „ „ : *q* and not *p*. $(q \,.\, \sim p)$
(*T F F F*) (p, q) „ „ : *q* and *p*. $(q \,.\, p)$
(*F F F F*) (p, q) Contradiction (*p* and not *p*, and *q* and not *q*.) $(p \,.\, \sim p \,.\, q \,.\, \sim q)$

I will give the name *truth-grounds* of a proposition to those truth-possibilities of its truth-arguments that make it true.

5.11 If all the truth-grounds that are common to a number of propositions are at the same time truth-grounds of a certain proposition, then we say that the truth of that proposition follows from the truth of the others.

5.12 In particular, the truth of a proposition '*p*' follows from the truth of another proposition '*q*' if all the truth-grounds of the latter are truth-grounds of the former.

5.121 The truth-grounds of the one are contained in those of the other: *p* follows from *q*.

5.122 If *p* follows from *q*, the sense of '*p*' is contained in the sense of '*q*'.

5.123 If a god creates a world in which certain propositions are true, then by that very act he also creates a world in which all the propositions that follow from them come true. And similarly he could not create a world in which the proposition '*p*' was true without creating all its objects.

5.124 A proposition affirms every proposition that follows from it.

5.1241 '*p*.*q*' is one of the propositions that affirm '*p*' and at the same time one of the propositions that affirm '*q*'.

Zwei Sätze sind einander entgegengesetzt, wenn es keinen sinnvollen Satz gibt, der sie beide bejaht.

Jeder Satz, der einem anderen widerspricht, verneint ihn.

5.13 Daß die Wahrheit eines Satzes aus der Wahrheit anderer Sätze folgt, ersehen wir aus der Struktur der Sätze.

5.131 Folgt die Wahrheit eines Satzes aus der Wahrheit anderer, so drückt sich dies durch Beziehungen aus, in welchen die Formen jener Sätze zu einander stehen; und zwar brauchen wir sie nicht erst in jene Beziehungen zu setzen, indem wir sie in einem Satze miteinander verbinden, sondern diese Beziehungen sind intern und bestehen, sobald, und dadurch daß, jene Sätze bestehen.

5.1311 Wenn wir von pvq und ~p auf q schließen, so ist hier durch die Bezeichnungsweise die Beziehung der Satzformen von „p v q" und „~p" verhüllt. Schreiben wir aber z. B. statt „p v q" „p|q.|.p|q", und statt „~p" „p|p" (p|q = weder p, noch q), so wird der innere Zusammenhang offenbar.

(Daß man aus (x).fx auf fa schließen kann, das zeigt, daß die Allgemeinheit auch im Symbol „(x).fx" vorhanden ist.)

5.132 Folgt p aus q, so kann ich von q auf p schließen; p aus q folgern.

Die Art des Schlusses ist allein aus den beiden Sätzen zu entnehmen.

Nur sie selbst können den Schluß rechtfertigen.

„Schlußgesetze", welche — wie bei Frege und Russell — die Schlüsse rechtfertigen sollen, sind sinnlos, und wären überflüssig.

5.133 Alles Folgern geschieht a priori.

5.134 Aus einem Elementarsatz läßt sich kein anderer folgern.

5.135 Auf keine Weise kann aus dem Bestehen irgend einer Sachlage auf das Bestehen einer von ihr gänzlich verschiedenen Sachlage geschlossen werden.

Two propositions are opposed to one another if there is no proposition with a sense, that affirms them both.

Every proposition that contradicts another negates it.

5.13 When the truth of one proposition follows from the truth of others, we can see this from the structure of the propositions.

5.131 If the truth of one proposition follows from the truth of others, this finds expression in relations in which the forms of the propositions stand to one another: nor is it necessary for us to set up these relations between them, by combining them with one another in a single proposition; on the contrary, the relations are internal, and their existence is an immediate result of the existence of the propositions.

5.1311 When we infer q from $p \vee q$ and $\sim p$, the relation between the propositional forms of '$p \vee q$' and '$\sim p$' is masked, in this case, by our mode of signifying. But if instead of '$p \vee q$' we write, for example, '$p|q . | . p|q$', and instead of '$\sim p$', '$p|p$' ($p|q$ = neither p nor q), then the inner connexion becomes obvious.

(The possibility of inference from $(x) . fx$ to fa shows that the symbol $(x) . fx$ itself has generality in it.)

5.132 If p follows from q, I can make an inference from q to p, deduce p from q.

The nature of the inference can be gathered only from the two propositions.

They themselves are the only possible justification of the inference.

'Laws of inference', which are supposed to justify inferences, as in the works of Frege and Russell, have no sense, and would be superfluous.

5.133 All deductions are made a priori.

5.134 One elementary proposition cannot be deduced from another.

5.135 There is no possible way of making an inference from the existence of one situation to the existence of another, entirely different situation.

5.136 Einen Kausalnexus, der einen solchen Schluß rechtfertigte, gibt es nicht.

5.1361 Die Ereignisse der Zukunft *können* wir nicht aus den gegenwärtigen erschließen.

Der Glaube an den Kausalnexus ist der *Aberglaube*.

5.1362 Die Willensfreiheit besteht darin, daß zukünftige Handlungen jetzt nicht gewußt werden können. Nur dann könnten wir sie wissen, wenn die Kausalität eine *innere* Notwendigkeit wäre, wie die des logischen Schlusses.— Der Zusammenhang von Wissen und Gewußtem ist der der logischen Notwendigkeit.

(„A weiß, daß p der Fall ist" ist sinnlos, wenn p eine Tautologie ist.)

5.1363 Wenn daraus, daß ein Satz uns einleuchtet, nicht *folgt*, daß er wahr ist, so ist das Einleuchten auch keine Rechtfertigung für unseren Glauben an seine Wahrheit.

5.14 Folgt ein Satz aus einem anderen, so sagt dieser mehr als jener, jener weniger als dieser.

5.141 Folgt p aus q und q aus p, so sind sie ein und derselbe Satz.

5.142 Die Tautologie folgt aus allen Sätzen: sie sagt nichts.

5.143 Die Kontradiktion ist das Gemeinsame der Sätze, was *kein* Satz mit einem anderen gemein hat. Die Tautologie ist das Gemeinsame aller Sätze, welche nichts miteinander gemein haben.

Die Kontradiktion verschwindet sozusagen außerhalb, die Tautologie innerhalb aller Sätze.

Die Kontradiktion ist die äußere Grenze der Sätze, die Tautologie ihr substanzloser Mittelpunkt.

5.15 Ist W_r die Anzahl der Wahrheitsgründe des Satzes „r", W_{rs} die Anzahl derjenigen Wahrheitsgründe des Satzes „s", die zugleich Wahrheitsgründe von „r" sind, dann nennen wir das Verhältnis: W_{rs}: W_r das Maß der *Wahrscheinlichkeit*, welche der Satz „r" dem Satz „s" gibt.

5.151 Sei in einem Schema wie dem obigen in No. 5.101 W_r die Anzahl der „W" im Satze r; W_{rs} die Anzahl derjenigen

5.136 There is no causal nexus to justify such an inference.

5.1361 We *cannot* infer the events of the future from those of the present.

Superstition is nothing but belief in the causal nexus.

5.1362 The freedom of the will consists in the impossibility of knowing actions that still lie in the future. We could know them only if causality were an *inner* necessity like that of logical inference.—The connexion between knowledge and what is known is that of logical necessity.

('*A* knows that p is the case', has no sense if p is a tautology.)

5.1363 If the truth of a proposition does not *follow* from the fact that it is self-evident to us, then its self-evidence in no way justifies our belief in its truth.

5.14 If one proposition follows from another, then the latter says more than the former, and the former less than the latter.

5.141 If p follows from q and q from p, then they are one and the same proposition.

5.142 A tautology follows from all propositions: it says nothing.

5.143 Contradiction is that common factor of propositions which *no* proposition has in common with another. Tautology is the common factor of all propositions that have nothing in common with one another.

Contradiction, one might say, vanishes outside all propositions: tautology vanishes inside them.

Contradiction is the outer limit of propositions: tautology is the unsubstantial point at their centre.

5.15 If T_r is the number of the truth-grounds of a proposition 'r', and if T_{rs} is the number of the truth-grounds of a proposition 's' that are at the same time truth-grounds of 'r', then we call the ratio $T_{rs} : T_r$ the degree of *probability* that the proposition 'r' gives to the proposition 's'.

5.151 In a schema like the one above in 5.101, let T_r be the number of 'T's' in the proposition r, and let T_{rs} be the

„W" im Satze s, die in gleichen Kolonnen mit „W" des Satzes r stehen. Der Satz r gibt dann dem Satze s die Wahrscheinlichkeit: $W_{rs} : W_r$.

5.1511 Es gibt keinen besonderen Gegenstand, der den Wahrscheinlichkeitssätzen eigen wäre.

5.152 Sätze, welche keine Wahrheitsargumente mit einander gemein haben, nennen wir von einander unabhängig.

Zwei Elementarsätze geben einander die Wahrscheinlichkeit $\frac{1}{2}$.

Folgt p aus q, so gibt der Satz „q" dem Satz „p" die Wahrscheinlichkeit 1. Die Gewißheit des logischen Schlusses ist ein Grenzfall der Wahrscheinlichkeit.

(Anwendung auf Tautologie und Kontradiktion.)

5.153 Ein Satz ist an sich weder wahrscheinlich noch unwahrscheinlich. Ein Ereignis trifft ein oder es trifft nicht ein, ein Mittelding gibt es nicht.

5.154 In einer Urne seien gleichviel weiße und schwarze Kugeln (und keine anderen). Ich ziehe eine Kugel nach der anderen und lege sie wieder in die Urne zurück. Dann kann ich durch den Versuch feststellen, daß sich die Zahlen der gezogenen schwarzen und weißen Kugeln bei fortgesetztem Ziehen einander nähern.

D a s ist also kein mathematisches Faktum.

Wenn ich nun sage: Es ist gleich wahrscheinlich, daß ich eine weiße Kugel wie eine schwarze ziehen werde, so heißt das: Alle mir bekannten Umstände (die hypothetisch angenommenen Naturgesetze mitinbegriffen) geben dem Eintreffen des einen Ereignisses nicht m e h r Wahrscheinlichkeit als dem Eintreffen des anderen. Das heißt, sie geben — wie aus den obigen Erklärungen leicht zu entnehmen ist — jedem die Wahrscheinlichkeit $\frac{1}{2}$.

Was ich durch den Versuch bestätige, ist, daß das Eintreffen der beiden Ereignisse von den Umständen, die ich nicht näher kenne, unabhängig ist.

5.155 Die Einheit des Wahrscheinlichkeitssatzes ist: Die Umstände — die ich sonst nicht weiter kenne — geben dem Eintreffen eines bestimmten Ereignisses den und den Grad der Wahrscheinlichkeit.

number of 'T's' in the proposition s that stand in columns in which the proposition r has 'T's'. Then the proposition r gives to the proposition s the probability $T_{rs} : T_r$.

5.1511 There is no special object peculiar to probability propositions.

5.152 When propositions have no truth-arguments in common with one another, we call them independent of one another.

Two elementary propositions give one another the probability $\frac{1}{2}$.

If p follows from q, then the proposition 'q' gives to the proposition 'p' the probability 1. The certainty of logical inference is a limiting case of probability.

(Application of this to tautology and contradiction.)

5.153 In itself, a proposition is neither probable nor improbable. Either an event occurs or it does not: there is no middle way.

5.154 Suppose that an urn contains black and white balls in equal numbers (and none of any other kind). I draw one ball after another, putting them back into the urn. By this experiment I can establish that the number of black balls drawn and the number of white balls drawn approximate to one another as the draw continues.

So *this* is not a mathematical truth.

Now, if I say, 'The probability of my drawing a white ball is equal to the probability of my drawing a black one', this means that all the circumstances that I know of (including the laws of nature assumed as hypotheses) give no *more* probability to the occurrence of the one event than to that of the other. That is to say, they give each the probability $\frac{1}{2}$, as can easily be gathered from the above definitions.

What I confirm by the experiment is that the occurrence of the two events is independent of the circumstances of which I have no more detailed knowledge.

5.155 The minimal unit for a probability proposition is this: The circumstances—of which I have no further knowledge—give such and such a degree of probability to the occurrence of a particular event.

5.156 So ist die Wahrscheinlichkeit eine Verallgemeinerung.

Sie involviert eine allgemeine Beschreibung einer Satzform.

Nur in Ermanglung der Gewißheit gebrauchen wir die Wahrscheinlichkeit.— Wenn wir zwar eine Tatsache nicht vollkommen kennen, wohl aber etwas über ihre Form wissen.

(Ein Satz kann zwar ein unvollständiges Bild einer gewissen Sachlage sein, aber er ist immer ein vollständiges Bild.)

Der Wahrscheinlichkeitssatz ist gleichsam ein Auszug aus anderen Sätzen.

5.2 Die Strukturen der Sätze stehen in internen Beziehungen zu einander.

5.21 Wir können diese internen Beziehungen dadurch in unserer Ausdrucksweise hervorheben, daß wir einen Satz als Resultat einer Operation darstellen, die ihn aus anderen Sätzen (den Basen der Operation) hervorbringt.

5.22 Die Operation ist der Ausdruck einer Beziehung zwischen den Strukturen ihres Resultats und ihrer Basen.

5.23 Die Operation ist das, was mit dem einen Satz geschehen muß, um aus ihm den anderen zu machen.

5.231 Und das wird natürlich von ihren formalen Eigenschaften, von der internen Ähnlichkeit ihrer Formen abhängen.

5.232 Die interne Relation, die eine Reihe ordnet, ist äquivalent mit der Operation, durch welche ein Glied aus dem anderen entsteht.

5.233 Die Operation kann erst dort auftreten, wo ein Satz auf logisch bedeutungsvolle Weise aus einem anderen entsteht. Also dort, wo die logische Konstruktion des Satzes anfängt.

5.234 Die Wahrheitsfunktionen der Elementarsätze sind Resultate von Operationen, die die Elementarsätze als Basen haben. (Ich nenne diese Operationen Wahrheitsoperationen.)

5.2341 Der Sinn einer Wahrheitsfunktion von p ist eine Funktion des Sinnes von p.

5.156 It is in this way that probability is a generalization.

It involves a general description of a propositional form.

We use probability only in default of certainty—if our knowledge of a fact is not indeed complete, but we do know *something* about its form.

(A proposition may well be an incomplete picture of a certain situation, but it is always a complete picture of *something*.)

A probability proposition is a sort of excerpt from other propositions.

5.2 The structures of propositions stand in internal relations to one another.

5.21 In order to give prominence to these internal relations we can adopt the following mode of expression: we can represent a proposition as the result of an operation that produces it out of other propositions (which are the bases of the operation).

5.22 An operation is the expression of a relation between the structures of its result and of its bases.

5.23 The operation is what has to be done to the one proposition in order to make the other out of it.

5.231 And that will, of course, depend on their formal properties, on the internal similarity of their forms.

5.232 The internal relation by which a series is ordered is equivalent to the operation that produces one term from another.

5.233 Operations cannot make their appearance before the point at which one proposition is generated out of another in a logically meaningful way; i.e. the point at which the logical construction of propositions begins.

5.234 Truth-functions of elementary propositions are results of operations with elementary propositions as bases. (These operations I call truth-operations.)

5.2341 The sense of a truth-function of p is a function of the sense of p.

Verneinung, logische Addition, logische Multiplikation, etc. etc. sind Operationen.

(Die Verneinung verkehrt den Sinn des Satzes.)

5.24 Die Operation zeigt sich in einer Variablen; sie zeigt, wie man von einer Form von Sätzen zu einer anderen gelangen kann.

Sie bringt den Unterschied der Formen zum Ausdruck.

(Und das Gemeinsame zwischen den Basen und dem Resultat der Operation sind eben die Basen.)

5.241 Die Operation kennzeichnet keine Form, sondern nur den Unterschied der Formen.

5.242 Dieselbe Operation, die „q“ aus „p“ macht, macht aus „q“ „r“, u. s. f. Dies kann nur darin ausgedrückt sein, daß „p“, „q“, „r“, etc. Variable sind, die gewisse formale Relationen allgemein zum Ausdruck bringen.

5.25 Das Vorkommen der Operation charakterisiert den Sinn des Satzes nicht.

Die Operation sagt ja nichts aus, nur ihr Resultat, und dies hängt von den Basen der Operation ab.

(Operation und Funktion dürfen nicht miteinander verwechselt werden.)

5.251 Eine Funktion kann nicht ihr eigenes Argument sein, wohl aber kann das Resultat einer Operation ihre eigene Basis werden.

5.252 Nur so ist das Fortschreiten von Glied zu Glied in einer Formenreihe (von Type zu Type in den Hierarchien Russells und Whiteheads) möglich. (Russell und Whitehead haben die Möglichkeit dieses Fortschreitens nicht zugegeben, aber immer wieder von ihr Gebrauch gemacht.)

5.2521 Die fortgesetzte Anwendung einer Operation auf ihr eigenes Resultat nenne ich ihre successive Anwendung („O’O’O’a“ ist das Resultat der dreimaligen successiven Anwendung von „O’ξ“ auf „a“).

In einem ähnlichen Sinne rede ich von der successiven Anwendung m e h r e r e r Operationen auf eine Anzahl von Sätzen.

Negation, logical addition, logical multiplication, etc. etc. are operations.

(Negation reverses the sense of a proposition.)

5.24 An operation manifests itself in a variable; it shows how we can get from one form of proposition to another.

It gives expression to the difference between the forms.

(And what the bases of an operation and its result have in common is just the bases themselves.)

5.241 An operation is not the mark of a form, but only of a difference between forms.

5.242 The operation that produces '*q*' from '*p*' also produces '*r*' from '*q*', and so on. There is only one way of expressing this: '*p*', '*q*', '*r*', etc. have to be variables that give expression in a general way to certain formal relations.

5.25 The occurrence of an operation does not characterize the sense of a proposition.

Indeed, no statement is made by an operation, but only by its result, and this depends on the bases of the operation.

(Operations and functions must not be confused with each other.)

5.251 A function cannot be its own argument, whereas an operation can take one of its own results as its base.

5.252 It is only in this way that the step from one term of a series of forms to another is possible (from one type to another in the hierarchies of Russell and Whitehead). (Russell and Whitehead did not admit the possibility of such steps, but repeatedly availed themselves of it.)

5.2521 If an operation is applied repeatedly to its own results, I speak of successive applications of it. ('*O*'*O*'*O*'*a*' is the result of three successive applications of the operation '*O*'*ξ*' to '*a*'.)

In a similar sense I speak of successive applications of *more than one* operation to a number of propositions.

5.2522 Das allgemeine Glied einer Formenreihe a, O'a, O'O'a, ... schreibe ich daher so: „[a, x, O'x]". Dieser Klammerausdruck ist eine Variable. Das erste Glied des Klammerausdruckes ist der Anfang der Formenreihe, das zweite die Form eines beliebigen Gliedes x der Reihe und das dritte die Form desjenigen Gliedes der Reihe, welches auf x unmittelbar folgt.

5.2523 Der Begriff der successiven Anwendung der Operation ist äquivalent mit dem Begriff „und so weiter".

5.253 Eine Operation kann die Wirkung einer anderen rückgängig machen. Operationen können einander aufheben.

5.254 Die Operation kann verschwinden (z. B. die Verneinung in „~~p": ~~p = p).

5.3 Alle Sätze sind Resultate von Wahrheitsoperationen mit den Elementarsätzen.

Die Wahrheitsoperation ist die Art und Weise, wie aus den Elementarsätzen die Wahrheitsfunktion entsteht.

Nach dem Wesen der Wahrheitsoperation wird auf die gleiche Weise, wie aus den Elementarsätzen ihre Wahrheitsfunktion, aus Wahrheitsfunktionen eine neue. Jede Wahrheitsoperation erzeugt aus Wahrheitsfunktionen von Elementarsätzen wieder eine Wahrheitsfunktion von Elementarsätzen, einen Satz. Das Resultat jeder Wahrheitsoperation mit den Resultaten von Wahrheitsoperationen mit Elementarsätzen ist wieder das Resultat Einer Wahrheitsoperation mit Elementarsätzen.

Jeder Satz ist das Resultat von Wahrheitsoperationen mit Elementarsätzen.

5.31 Die Schemata No. 4.31 haben auch dann eine Bedeutung, wenn „p", „q", „r", etc. nicht Elementarsätze sind.

Und es ist leicht zu sehen, daß das Satzzeichen in No. 4.442, auch wenn „p" und „q" Wahrheitsfunktionen von Elementarsätzen sind, Eine Wahrheitsfunktion von Elementarsätzen ausdrückt.

5.32 Alle Wahrheitsfunktionen sind Resultate der successiven Anwendung einer endlichen Anzahl von Wahrheitsoperationen auf die Elementarsätze.

5.2522 Accordingly I use the sign '[*a*, *x*, *O'x*]' for the general term of the series of forms *a*, *O'a*, *O'O'a*, ... This bracketed expression is a variable: the first term of the bracketed expression is the beginning of the series of forms, the second is the form of a term *x* arbitrarily selected from the series, and the third is the form of the term that immediately follows *x* in the series.

5.2523 The concept of successive applications of an operation is equivalent to the concept 'and so on'.

5.253 One operation can counteract the effect of another. Operations can cancel one another.

5.254 An operation can vanish (e.g. negation in '$\sim\sim p$': $\sim\sim p = p$).

5.3 All propositions are results of truth-operations on elementary propositions.

A truth-operation is the way in which a truth-function is produced out of elementary propositions.

It is of the essence of truth-operations that, just as elementary propositions yield a truth-function of themselves, so too in the same way truth-functions yield a further truth-function. When a truth-operation is applied to truth-functions of elementary propositions, it always generates another truth-function of elementary propositions, another proposition. When a truth-operation is applied to the results of truth-operations on elementary propositions, there is always a *single* operation on elementary propositions that has the same result.

Every proposition is the result of truth-operations on elementary propositions.

5.31 The schemata in 4.31 have a meaning even when '*p*', '*q*', '*r*', etc. are not elementary propositions.

And it is easy to see that the propositional sign in 4.442 expresses a single truth-function of elementary propositions even when '*p*' and '*q*' are truth-functions of elementary propositions.

5.32 All truth-functions are results of successive applications to elementary propositions of a finite number of truth-operations.

5.4 Hier zeigt es sich, daß es „logische Gegenstände", „logische Konstante" (im Sinne Freges und Russells) nicht gibt.

5.41 Denn: Alle Resultate von Wahrheitsoperationen mit Wahrheitsfunktionen sind identisch, welche eine und dieselbe Wahrheitsfunktion von Elementarsätzen sind.

5.42 Dass v, ⊃, etc. nicht Beziehungen im Sinne von rechts und links etc. sind, leuchtet ein.

Die Möglichkeit des kreuzweisen Definierens der logischen „Urzeichen" Freges und Russells zeigt schon, daß diese keine Urzeichen sind, und schon erst recht, daß sie keine Relationen bezeichnen.

Und es ist offenbar, daß das „⊃", welches wir durch „~" und „v" definieren, identisch ist mit dem, durch welches wir „v" mit „~" definieren, und daß dieses „v" mit dem ersten identisch ist. U. s. w.

5.43 Daß aus einer Tatsache p unendlich viele a n d e r e folgen sollten, nämlich ~~p, ~~~~p, etc., ist doch von vornherein kaum zu glauben. Und nicht weniger merkwürdig ist, daß die unendliche Anzahl der Sätze der Logik (der Mathematik) aus einem halben Dutzend „Grundgesetzen" folgen.

Alle Sätze der Logik sagen aber dasselbe. Nämlich nichts.

5.44 Die Wahrheitsfunktionen sind keine materiellen Funktionen.

Wenn man z. B. eine Bejahung durch doppelte Verneinung erzeugen kann, ist dann die Verneinung — in irgend einem Sinn — in der Bejahung enthalten? Verneint „~~p" ~p, oder bejaht es p; oder beides?

Der Satz „~~p" handelt nicht von der Verneinung wie von einem Gegenstand; wohl aber ist die Möglichkeit der Verneinung in der Bejahung bereits präjudiziert.

Und gäbe es einen Gegenstand, der „~" hieße, so müßte „~~p" etwas anderes sagen als „p". Denn der eine Satz würde dann eben von ~ handeln, der andere nicht.

5.4 At this point it becomes manifest that there are no 'logical objects' or 'logical constants' (in Frege's and Russell's sense).

5.41 The reason is that the results of truth-operations on truth-functions are always identical whenever they are one and the same truth-function of elementary propositions.

5.42 It is self-evident that v, ⊃, etc. are not relations in the sense in which right and left etc. are relations.

The interdefinability of Frege's and Russell's 'primitive signs' of logic is enough to show that they are not primitive signs, still less signs for relations.

And it is obvious that the '⊃' defined by means of '~' and 'v' is identical with the one that figures with '~' in the definition of 'v'; and that the second 'v' is identical with the first one; and so on.

5.43 Even at first sight it seems scarcely credible that there should follow from one fact *p* infinitely many *others*, namely ~~*p*, ~~~~*p*, etc. And it is no less remarkable that the infinite number of propositions of logic (mathematics) follow from half a dozen 'primitive propositions'.

But in fact all the propositions of logic say the same thing, to wit nothing.

5.44 Truth-functions are not material functions.

For example, an affirmation can be produced by double negation: in such a case does it follow that in some sense negation is contained in affirmation? Does '~~*p*' negate ~*p*, or does it affirm *p*—or both?

The proposition '~~*p*' is not about negation, as if negation were an object: on the other hand, the possibility of negation is already written into affirmation.

And if there were an object called '~', it would follow that '~~*p*' said something different from what '*p*' said, just because the one proposition would then be about ~ and the other would not.

5.441 Dieses Verschwinden der scheinbaren logischen Konstanten tritt auch ein, wenn „~(∃x).~fx" dasselbe sagt wie „(x).fx", oder „(∃x).fx.x = a" dasselbe wie „fa".

5.442 Wenn uns ein Satz gegeben ist, so sind mit ihm auch schon die Resultate aller Wahrheitsoperationen, die ihn zur Basis haben, gegeben.

5.45 Gibt es logische Urzeichen, so muß eine richtige Logik ihre Stellung zueinander klar machen und ihr Dasein rechtfertigen. Der Bau der Logik aus ihren Urzeichen muß klar werden.

5.451 Hat die Logik Grundbegriffe, so müssen sie von einander unabhängig sein. Ist ein Grundbegriff eingeführt, so muß er in allen Verbindungen eingeführt sein, worin er überhaupt vorkommt. Man kann ihn also nicht zuerst für eine Verbindung, dann noch einmal für eine andere einführen. Z. B.: Ist die Verneinung eingeführt, so müssen wir sie jetzt in Sätzen von der Form „~p" ebenso verstehen, wie in Sätzen wie „~(p v q)", „(∃x).~fx", u. a. Wir dürfen sie nicht erst für die eine Klasse von Fällen, dann für die andere einführen, denn es bliebe dann zweifelhaft, ob ihre Bedeutung in beiden Fällen die gleiche wäre und es wäre kein Grund vorhanden, in beiden Fällen dieselbe Art der Zeichenverbindung zu benützen.

(Kurz, für die Einführung der Urzeichen gilt, mutatis mutandis, dasselbe, was Frege („Grundgesetze der Arithmetik") für die Einführung von Zeichen durch Definitionen gesagt hat.)

5.452 Die Einführung eines neuen Behelfes in den Symbolismus der Logik muß immer ein folgenschweres Ereignis sein. Kein neuer Behelf darf in die Logik — sozusagen, mit ganz unschuldiger Miene — in Klammern oder unter dem Striche eingeführt werden.

(So kommen in den „Principia Mathematica" von Russell und Whitehead Definitionen und Grundgesetze in Worten vor. Warum hier plötzlich Worte? Dies bedürfte einer Rechtfertigung. Sie fehlt und muß fehlen, da das Vorgehen tatsächlich unerlaubt ist.)

5.441 This vanishing of the apparent logical constants also occurs in the case of '$\sim(\exists x) . \sim fx$', which says the same as '$(x) . fx$', and in the case of '$(\exists x) . fx . x = a$', which says the same as 'fa'.

5.442 If we are given a proposition, then *with it* we are also given the results of all truth-operations that have it as their base.

5.45 If there are primitive logical signs, then any logic that fails to show clearly how they are placed relatively to one another and to justify their existence will be incorrect. The construction of logic *out of* its primitive signs must be made clear.

5.451 If logic has primitive ideas, they must be independent of one another. If a primitive idea has been introduced, it must have been introduced in all the combinations in which it ever occurs. It cannot, therefore, be introduced first for *one* combination and later re-introduced for another. For example, once negation has been introduced, we must understand it both in propositions of the form '$\sim p$' and in propositions like '$\sim(p \vee q)$', '$(\exists x) . \sim fx$', etc. We must not introduce it first for the one class of cases and then for the other, since it would then be left in doubt whether its meaning were the same in both cases, and no reason would have been given for combining the signs in the same way in both cases.

(In short, Frege's remarks about introducing signs by means of definitions (in *The Fundamental Laws of Arithmetic*) also apply, mutatis mutandis, to the introduction of primitive signs.)

5.452 The introduction of any new device into the symbolism of logic is necessarily a momentous event. In logic a new device should not be introduced in brackets or in a footnote with what one might call a completely innocent air.

(Thus in Russell and Whitehead's *Principia Mathematica* there occur definitions and primitive propositions expressed in words. Why this sudden appearance of words? It would require a justification, but none is given, or could be given, since the procedure is in fact illicit.)

Hat sich aber die Einführung eines neuen Behelfes an einer Stelle als nötig erwiesen, so muß man sich nun sofort fragen: Wo muß dieser Behelf nun *immer* angewandt werden? Seine Stellung in der Logik muß nun erklärt werden.

5.453 Alle Zahlen der Logik müssen sich rechtfertigen lassen.

Oder vielmehr: Es muß sich herausstellen, daß es in der Logik keine Zahlen gibt.

Es gibt keine ausgezeichneten Zahlen.

5.454 In der Logik gibt es kein Nebeneinander, kann es keine Klassifikation geben.

In der Logik kann es nicht Allgemeineres und Spezielleres geben.

5.4541 Die Lösungen der logischen Probleme müssen einfach sein, denn sie setzen den Standard der Einfachheit.

Die Menschen haben immer geahnt, daß es ein Gebiet von Fragen geben müsse, deren Antworten — a priori — symmetrisch, und zu einem abgeschlossenen, regelmäßigen Gebilde vereint liegen.

Ein Gebiet, in dem der Satz gilt: Simplex sigillum veri.

5.46 Wenn man die logischen Zeichen richtig einführte, so hätte man damit auch schon den Sinn aller ihrer Kombinationen eingeführt; also nicht nur „p v q“ sondern auch schon „~(p v ~q)“, etc. etc. Man hätte damit auch schon die Wirkung aller nur möglichen Kombinationen von Klammern eingeführt. Und damit wäre es klar geworden, daß die eigentlichen allgemeinen Urzeichen nicht die „p v q“, „(∃x).fx“, etc. sind, sondern die allgemeinste Form ihrer Kombinationen.

5.461 Bedeutungsvoll ist die scheinbar unwichtige Tatsache, daß die logischen Scheinbeziehungen, wie v und ⊃, der Klammern bedürfen — im Gegensatz zu den wirklichen Beziehungen.

Die Benützung der Klammern mit jenen scheinbaren Urzeichen deutet ja schon darauf hin, daß diese nicht die wirklichen Urzeichen sind. Und es wird doch wohl niemand glauben, daß die Klammern eine selbständige Bedeutung haben.

But if the introduction of a new device has proved necessary at a certain point, we must immediately ask ourselves, 'At what points is the employment of this device now *unavoidable*?' and its place in logic must be made clear.

5.453 All numbers in logic stand in need of justification.

Or rather, it must become evident that there are no numbers in logic.

There are no pre-eminent numbers.

5.454 In logic there is no co-ordinate status, and there can be no classification.

In logic there can be no distinction between the general and the specific.

5.4541 The solutions of the problems of logic must be simple, since they set the standard of simplicity.

Men have always had a presentiment that there must be a realm in which the answers to questions are symmetrically combined—a priori—to form a self-contained system.

A realm subject to the law: Simplex sigillum veri.

5.46 If we introduced logical signs properly, then we should also have introduced at the same time the sense of all combinations of them; i.e. not only '$p \vee q$' but '$\sim(p \vee \sim q)$' as well, etc. etc. We should also have introduced at the same time the effect of all possible combinations of brackets. And thus it would have been made clear that the real general primitive signs are not '$p \vee q$', '$(\exists x).fx$', etc. but the most general form of their combinations.

5.461 Though it seems unimportant, it is in fact significant that the pseudo-relations of logic, such as $\vee$ and $\supset$, need brackets—unlike real relations.

Indeed, the use of brackets with these apparently primitive signs is itself an indication that they are not the real primitive signs. And surely no one is going to believe that brackets have an independent meaning.

5.4611 Die logischen Operationszeichen sind Interpunktionen.

5.47 Es ist klar, daß alles, was sich überhaupt von vornherein über die Form aller Sätze sagen läßt, sich auf einmal sagen lassen muß.

Sind ja schon im Elementarsatze alle logischen Operationen enthalten. Denn „fa" sagt dasselbe wie

$$„(\exists x).fx.x = a".$$

Wo Zusammengesetztheit ist, da ist Argument und Funktion, und wo diese sind, sind bereits alle logischen Konstanten.

Man könnte sagen: Die Eine logische Konstante ist das, was alle Sätze, ihrer Natur nach, mit einander gemein haben.

Das aber ist die allgemeine Satzform.

5.471 Die allgemeine Satzform ist das Wesen des Satzes.

5.4711 Das Wesen des Satzes angeben heißt das Wesen aller Beschreibung angeben, also das Wesen der Welt.

5.472 Die Beschreibung der allgemeinsten Satzform ist die Beschreibung des einen und einzigen allgemeinen Urzeichens der Logik.

5.473 Die Logik muß für sich selber sorgen.

Ein mögliches Zeichen muß auch bezeichnen können. Alles was in der Logik möglich ist, ist auch erlaubt. („Sokrates ist identisch", heißt darum nichts, weil es keine Eigenschaft gibt, die „identisch" heißt. Der Satz ist unsinnig, weil wir eine willkürliche Bestimmung nicht getroffen haben, aber nicht darum, weil das Symbol an und für sich unerlaubt wäre.)

Wir können uns, in gewissem Sinne, nicht in der Logik irren.

5.4731 Das Einleuchten, von dem Russell so viel sprach, kann nur dadurch in der Logik entbehrlich werden, daß die Sprache selbst jeden logischen Fehler verhindert.— Daß die Logik a priori ist, besteht darin, daß nicht unlogisch gedacht werden kann.

5.4611 Signs for logical operations are punctuation-marks.

5.47 It is clear that whatever we can say *in advance* about the form of all propositions, we must be able to say *all at once*.

An elementary proposition really contains all logical operations in itself. For 'fa' says the same thing as

$$`(\exists x).fx.x = a'.$$

Wherever there is compositeness, argument and function are present, and where these are present, we already have all the logical constants.

One could say that the sole logical constant was what *all* propositions, by their very nature, had in common with one another.

But that is the general propositional form.

5.471 The general propositional form is the essence of a proposition.

5.4711 To give the essence of a proposition means to give the essence of all description, and thus the essence of the world.

5.472 The description of the most general propositional form is the description of the one and only general primitive sign in logic.

5.473 Logic must look after itself.

If a sign is *possible*, then it is also capable of signifying. Whatever is possible in logic is also permitted. (The reason why 'Socrates is identical' means nothing is that there is no property called 'identical'. The proposition is nonsensical because we have failed to make an arbitrary determination, and not because the symbol, in itself, would be illegitimate.)

In a certain sense, we cannot make mistakes in logic.

5.4731 Self-evidence, which Russell talked about so much, can become dispensable in logic, only because language itself prevents every logical mistake.—What makes logic a priori is the *impossibility* of illogical thought.

5.4732 Wir können einem Zeichen nicht den unrechten Sinn geben.

5.47321 Occams Devise ist natürlich keine willkürliche, oder durch ihren praktischen Erfolg gerechtfertigte, Regel: Sie besagt, daß *unnötige* Zeicheneinheiten nichts bedeuten.

Zeichen, die *Einen* Zweck erfüllen, sind logisch äquivalent, Zeichen, die *keinen* Zweck erfüllen, logisch bedeutungslos.

5.4733 Frege sagt: Jeder rechtmäßig gebildete Satz muß einen Sinn haben; und ich sage: Jeder mögliche Satz ist rechtmäßig gebildet, und wenn er keinen Sinn hat, so kann das nur daran liegen, daß wir einigen seiner Bestandteile keine *Bedeutung* gegeben haben.

(Wenn wir auch glauben, es getan zu haben.)

So sagt „Sokrates ist identisch" darum nichts, weil wir dem Wort „identisch" als *Eigenschaftswort* *keine* Bedeutung gegeben haben. Denn, wenn es als Gleichheitszeichen auftritt, so symbolisiert es auf ganz andere Art und Weise — die bezeichnende Beziehung ist eine andere —, also ist auch das Symbol in beiden Fällen ganz verschieden; die beiden Symbole haben nur das Zeichen zufällig miteinander gemein.

5.474 Die Anzahl der nötigen Grundoperationen hängt *nur* von unserer Notation ab.

5.475 Es kommt nur darauf an, ein Zeichensystem von einer bestimmten Anzahl von Dimensionen — von einer bestimmten mathematischen Mannigfaltigkeit — zu bilden.

5.476 Es ist klar, daß es sich hier nicht um eine *Anzahl von Grundbegriffen* handelt, die bezeichnet werden müssen, sondern um den Ausdruck einer Regel.

5.5 Jede Wahrheitsfunktion ist ein Resultat der successiven Anwendung der Operation

$$(----\mathrm{W})(\xi, ...)$$

auf Elementarsätze.

Diese Operation verneint sämtliche Sätze in der rechten Klammer und ich nenne sie die Negation dieser Sätze.

5.4732 We cannot give a sign the wrong sense.

5.47321 Occam's maxim is, of course, not an arbitrary rule, nor one that is justified by its success in practice: its point is that *unnecessary* units in a sign-language mean nothing.

Signs that serve *one* purpose are logically equivalent, and signs that serve *none* are logically meaningless.

5.4733 Frege says that any legitimately constructed proposition must have a sense. And I say that any possible proposition is legitimately constructed, and, if it has no sense, that can only be because we have failed to give a *meaning* to some of its constituents.

(Even if we think that we have done so.)

Thus the reason why 'Socrates is identical' says nothing is that we have not given *any adjectival* meaning to the word 'identical'. For when it appears as a sign for identity, it symbolizes in an entirely different way—the signifying relation is a different one—therefore the symbols also are entirely different in the two cases: the two symbols have only the sign in common, and that is an accident.

5.474 The number of fundamental operations that are necessary depends *solely* on our notation.

5.475 All that is required is that we should construct a system of signs with a particular number of dimensions—with a particular mathematical multiplicity.

5.476 It is clear that this is not a question of a *number of primitive ideas* that have to be signified, but rather of the expression of a rule.

5.5 Every truth-function is a result of successive applications to elementary propositions of the operation

$$`(-----T)(\xi, ...)'.$$

This operation negates all the propositions in the right-hand pair of brackets, and I call it the negation of those propositions.

5.501 Einen Klammerausdruck, dessen Glieder Sätze sind, deute ich — wenn die Reihenfolge der Glieder in der Klammer gleichgültig ist — durch ein Zeichen von der Form „$(\bar{\xi})$" an. „ξ" ist eine Variable, deren Werte die Glieder des Klammerausdruckes sind; und der Strich über der Variablen deutet an, daß sie ihre sämtlichen Werte in der Klammer vertritt.

(Hat also ξ etwa die 3 Werte P, Q, R, so ist

$$(\bar{\xi}) = (P, Q, R).)$$

Die Werte der Variablen werden festgesetzt.

Die Festsetzung ist die Beschreibung der Sätze, welche die Variable vertritt.

Wie die Beschreibung der Glieder des Klammerausdruckes geschieht, ist unwesentlich.

Wir können drei Arten der Beschreibung unterscheiden: 1. Die direkte Aufzählung. In diesem Fall können wir statt der Variablen einfach ihre konstanten Werte setzen. 2. Die Angabe einer Funktion fx, deren Werte für alle Werte von x die zu beschreibenden Sätze sind. 3. Die Angabe eines formalen Gesetzes, nach welchem jene Sätze gebildet sind. In diesem Falle sind die Glieder des Klammerausdrucks sämtliche Glieder einer Formenreihe.

5.502 Ich schreibe also statt „$(-----W)(\xi, ...)$" „$N(\bar{\xi})$".

$N(\bar{\xi})$ ist die Negation sämtlicher Werte der Satzvariablen ξ.

5.503 Da sich offenbar leicht ausdrücken läßt, wie mit dieser Operation Sätze gebildet werden können und wie Sätze mit ihr nicht zu bilden sind, so muß dies auch einen exakten Ausdruck finden können.

5.51 Hat ξ nur einen Wert, so ist $N(\bar{\xi}) = \sim p$ (nicht p), hat es zwei Werte, so ist $N(\bar{\xi}) = \sim p . \sim q$ (weder p, noch q).

5.511 Wie kann die allumfassende, weltspiegelnde Logik so spezielle Haken und Manipulationen gebrauchen? Nur, indem sich alle diese zu einem unendlich feinen Netzwerk, zu dem großen Spiegel, verknüpfen.

5.512 „$\sim$p" ist wahr, wenn „p" falsch ist. Also in dem wahren Satz „$\sim$p" ist „p" ein falscher Satz. Wie kann ihn nun der Strich „$\sim$" mit der Wirklichkeit zum Stimmen bringen?

5.501 When a bracketed expression has propositions as its terms—and the order of the terms inside the brackets is indifferent—then I indicate it by a sign of the form '$(\bar{\xi})$'. 'ξ' is a variable whose values are terms of the bracketed expression and the bar over the variable indicates that it is the representative of all its values in the brackets.

(E.g. if ξ has the three values P, Q, R, then

$$(\bar{\xi}) = (P, Q, R).)$$

What the values of the variable are is something that is stipulated.

The stipulation is a description of the propositions that have the variable as their representative.

How the description of the terms of the bracketed expression is produced is not essential.

We *can* distinguish three kinds of description: 1. direct enumeration, in which case we can simply substitute for the variable the constants that are its values; 2. giving a function fx whose values for all values of x are the propositions to be described; 3. giving a formal law that governs the construction of the propositions, in which case the bracketed expression has as its members all the terms of a series of forms.

5.502 So instead of '$(-----T)(\xi, ...)$', I write '$N(\bar{\xi})$'.

$N(\bar{\xi})$ is the negation of all the values of the propositional variable ξ.

5.503 It is obvious that we can easily express how propositions may be constructed with this operation, and how they may not be constructed with it; so it must be possible to find an exact expression for this.

5.51 If ξ has only one value, then $N(\bar{\xi}) = \sim p$ (not p); if it has two values, then $N(\bar{\xi}) = \sim p . \sim q$ (neither p nor q).

5.511 How can logic—all-embracing logic, which mirrors the world—use such peculiar crotchets and contrivances? Only because they are all connected with one another in an infinitely fine network, the great mirror.

5.512 '$\sim p$' is true if 'p' is false. Therefore, in the proposition '$\sim p$', when it is true, 'p' is a false proposition. How then can the stroke '$\sim$' make it agree with reality?

Das, was in „~p“ verneint, ist aber nicht das „~“, sondern dasjenige, was allen Zeichen dieser Notation, welche p verneinen, gemeinsam ist.

Also die gemeinsame Regel, nach welcher „~p“, „~~~p“, „~p v ~p“, „~p . ~p“, etc. etc. (ad inf.) gebildet werden. Und dies Gemeinsame spiegelt die Verneinung wieder.

5.513 Man könnte sagen: Das Gemeinsame aller Symbole, die sowohl p als q bejahen, ist der Satz „p . q“. Das Gemeinsame aller Symbole, die entweder p oder q bejahen, ist der Satz „p v q“.

Und so kann man sagen: Zwei Sätze sind einander entgegengesetzt, wenn sie nichts miteinander gemein haben, und: Jeder Satz hat nur ein Negativ, weil es nur einen Satz gibt, der ganz außerhalb seiner liegt.

Es zeigt sich so auch in Russells Notation, daß „q : p v ~p“ dasselbe sagt wie „q“; daß „p v ~p“ nichts sagt.

5.514 Ist eine Notation festgelegt, so gibt es in ihr eine Regel, nach der alle p verneinenden Sätze gebildet werden, eine Regel, nach der alle p bejahenden Sätze gebildet werden, eine Regel, nach der alle p oder q bejahenden Sätze gebildet werden, u. s. f. Diese Regeln sind den Symbolen äquivalent und in ihnen spiegelt sich ihr Sinn wieder.

5.515 Es muß sich an unseren Symbolen zeigen, daß das, was durch „v“, „.“, etc. miteinander verbunden ist, Sätze sein müssen.

Und dies ist auch der Fall, denn das Symbol „p“ und „q“ setzt ja selbst das „v“, „~“, etc. voraus. Wenn das Zeichen „p“ in „p v q“ nicht für ein komplexes Zeichen steht, dann kann es allein nicht Sinn haben; dann können aber auch die mit „p“ gleichsinnigen Zeichen „p v p“, „p . p“, etc. keinen Sinn haben. Wenn aber „p v p“ keinen Sinn hat, dann kann auch „p v q“ keinen Sinn haben.

5.5151 Muß das Zeichen des negativen Satzes mit dem Zeichen des positiven gebildet werden? Warum sollte man den negativen Satz nicht durch eine negative Tatsache

But in '$\sim p$' it is not '$\sim$' that negates; it is rather what is common to all the signs of this notation that negate *p*.

That is to say the common rule that governs the construction of '$\sim p$', '$\sim\sim\sim p$', '$\sim p \vee \sim p$', '$\sim p . \sim p$', etc. etc. (ad inf.). And this common factor mirrors negation.

5.513 We might say that what is common to all symbols that affirm both *p* and *q* is the proposition '$p . q$'; and that what is common to all symbols that affirm either *p* or *q* is the proposition '$p \vee q$'.

And similarly we can say that two propositions are opposed to one another if they have nothing in common with one another, and that every proposition has only one negative, since there is only one proposition that lies completely outside it.

Thus in Russell's notation too it is manifest that '$q : p \vee \sim p$' says the same thing as 'q', that '$p \vee \sim p$' says nothing.

5.514 Once a notation has been established, there will be in it a rule governing the construction of all propositions that negate *p*, a rule governing the construction of all propositions that affirm *p*, and a rule governing the construction of all propositions that affirm *p* or *q*; and so on. These rules are equivalent to the symbols; and in them their sense is mirrored.

5.515 It must be manifest in our symbols that it can only be propositions that are combined with one another by 'v', '.', etc.

And this is indeed the case, since the symbol in 'p' and 'q' itself presupposes 'v', '$\sim$', etc. If the sign 'p' in '$p \vee q$' does not stand for a complex sign, then it cannot have sense by itself: but in that case the signs '$p \vee p$', '$p . p$', etc., which have the same sense as *p*, must also lack sense. But if '$p \vee p$' has no sense, then '$p \vee q$' cannot have a sense either.

5.5151 Must the sign of a negative proposition be constructed with that of the positive proposition? Why should it not be possible to express a negative proposition by means of

ausdrücken können. (Etwa: Wenn „a“ nicht in einer bestimmten Beziehung zu „b“ steht, könnte das ausdrücken, daß aRb nicht der Fall ist.)

Aber auch hier ist ja der negative Satz indirekt durch den positiven gebildet.

Der positive *Satz* muß die Existenz des negativen *Satzes* voraussetzen und umgekehrt.

5.52 Sind die Werte von ξ sämtliche Werte einer Funktion fx für alle Werte von x, so wird $N(\bar{\xi}) = \sim (\exists x).fx$.

5.521 Ich trenne den Begriff *Alle* von der Wahrheitsfunktion.

Frege und Russell haben die Allgemeinheit in Verbindung mit dem logischen Produkt oder der logischen Summe eingeführt. So wurde es schwer, die Sätze „$(\exists x).fx$“ und „$(x).fx$“, in welchen beide Ideen beschlossen liegen, zu verstehen.

5.522 Das Eigentümliche der Allgemeinheitsbezeichnung ist erstens, daß sie auf ein logisches Urbild hinweist, und zweitens, daß sie Konstante hervorhebt.

5.523 Die Allgemeinheitsbezeichnung tritt als Argument auf.

5.524 Wenn die Gegenstände gegeben sind, so sind uns damit auch schon *alle* Gegenstände gegeben.

Wenn die Elementarsätze gegeben sind, so sind damit auch *alle* Elementarsätze gegeben.

5.525 Es ist unrichtig, den Satz „$(\exists x).fx$“ — wie Russell dies tut — in Worten durch „fx ist *möglich*“ wiederzugeben.

Gewißheit, Möglichkeit oder Unmöglichkeit einer Sachlage wird nicht durch einen Satz ausgedrückt, sondern dadurch, daß ein Ausdruck eine Tautologie, ein sinnvoller Satz oder eine Kontradiktion ist.

Jener Präzedenzfall, auf den man sich immer berufen möchte, muß schon im Symbol selber liegen.

5.526 Man kann die Welt vollständig durch vollkommen verallgemeinerte Sätze beschreiben, das heißt also, ohne irgend einen Namen von vornherein einem bestimmten Gegenstand zuzuordnen.

a negative fact? (E.g. suppose that '*a*' does not stand in a certain relation to '*b*'; then this might be used to say that *aRb* was not the case.)

But really even in this case the negative proposition is constructed by an indirect use of the positive.

The positive ***proposition*** necessarily presupposes the existence of the negative ***proposition*** and vice versa.

5.52 If ξ has as its values all the values of a function fx for all values of x, then $N(\bar{\xi}) = \sim(\exists x).fx$.

5.521 I dissociate the concept *all* from truth-functions.

Frege and Russell introduced generality in association with logical product or logical sum. This made it difficult to understand the propositions '$(\exists x).fx$' and '$(x).fx$', in which both ideas are embedded.

5.522 What is peculiar to the generality-sign is first, that it indicates a logical prototype, and secondly, that it gives prominence to constants.

5.523 The generality-sign occurs as an argument.

5.524 If objects are given, then at the same time we are given *all* objects.

If elementary propositions are given, then at the same time *all* elementary propositions are given.

5.525 It is incorrect to render the proposition '$(\exists x).fx$' in the words, '*fx* is *possible*', as Russell does.

The certainty, possibility, or impossibility of a situation is not expressed by a proposition, but by an expression's being a tautology, a proposition with sense, or a contradiction.

The precedent to which we are constantly inclined to appeal must reside in the symbol itself.

5.526 We can describe the world completely by means of fully generalized propositions, i.e. without first correlating any name with a particular object.

Um dann auf die gewöhnliche Ausdrucksweise zu kommen, muß man einfach nach einem Ausdruck: „Es gibt ein und nur ein x, welches . . .", sagen: Und dies x ist a.

5.5261 Ein vollkommen verallgemeinerter Satz ist, wie jeder andere Satz, zusammengesetzt. (Dies zeigt sich daran, daß wir in „$(\exists x, \phi) . \phi x$" „$\phi$" und „x" getrennt erwähnen müssen. Beide stehen unabhängig in bezeichnenden Beziehungen zur Welt, wie im unverallgemeinerten Satz.)

Kennzeichen des zusammengesetzten Symbols: Es hat etwas mit a n d e r e n Symbolen gemeinsam.

5.5262 Es verändert ja die Wahr- oder Falschheit j e d e s Satzes etwas am allgemeinen Bau der Welt. Und der Spielraum, welcher ihrem Bau durch die Gesamtheit der Elementarsätze gelassen wird, ist eben derjenige, welchen die ganz allgemeinen Sätze begrenzen.

(Wenn ein Elementarsatz wahr ist, so ist damit doch jedenfalls Ein Elementarsatz m e h r wahr.)

5.53 Gleichheit des Gegenstandes drücke ich durch Gleichheit des Zeichens aus, und nicht mit Hilfe eines Gleichheitszeichens. Verschiedenheit der Gegenstände durch Verschiedenheit der Zeichen.

5.5301 Daß die Identität keine Relation zwischen Gegenständen ist, leuchtet ein. Dies wird sehr klar, wenn man z. B. den Satz „$(x):fx . \supset . x = a$" betrachtet. Was dieser Satz sagt, ist einfach, daß n u r a der Funktion f genügt, und nicht, daß nur solche Dinge der Funktion f genügen, welche eine gewisse Beziehung zu a haben.

Man könnte nun freilich sagen, daß eben n u r a diese Beziehung zu a habe, aber, um dies auszudrücken, brauchten wir das Gleichheitszeichen selber.

5.5302 Russells Definition von „=" genügt nicht; weil man nach ihr nicht sagen kann, daß zwei Gegenstände alle Eigenschaften gemeinsam haben. (Selbst wenn dieser Satz nie richtig ist, hat er doch S i n n.)

5.5303 Beiläufig gesprochen: Von z w e i Dingen zu sagen, sie seien identisch, ist ein Unsinn, und von E i n e m zu sagen, es sei identisch mit sich selbst, sagt gar nichts.

Then, in order to arrive at the customary mode of expression, we simply need to add, after an expression like, 'There is one and only one x such that . . .', the words, 'and that x is a'.

5.5261 A fully generalized proposition, like every other proposition, is composite. (This is shown by the fact that in '$(\exists x, \phi) . \phi x$' we have to mention '$\phi$' and '$x$' separately. They both, independently, stand in signifying relations to the world, just as is the case in ungeneralized propositions.)

It is a mark of a composite symbol that it has something in common with *other* symbols.

5.5262 The truth or falsity of *every* proposition does make some alteration in the general construction of the world. And the range that the totality of elementary propositions leaves open for its construction is exactly the same as that which is delimited by entirely general propositions.

(If an elementary proposition is true, that means, at any rate, *one more* true elementary proposition.)

5.53 Identity of object I express by identity of sign, and not by using a sign for identity. Difference of objects I express by difference of signs.

5.5301 It is self-evident that identity is not a relation between objects. This becomes very clear if one considers, for example, the proposition '$(x):fx.\supset.x = a$'. What this proposition says is simply that *only a* satisfies the function f, and not that only things that have a certain relation to a satisfy the function f.

Of course, it might then be said that *only a* did have this relation to a; but in order to express that, we should need the identity-sign itself.

5.5302 Russell's definition of '$=$' is inadequate, because according to it we cannot say that two objects have all their properties in common. (Even if this proposition is never correct, it still has *sense*.)

5.5303 Roughly speaking, to say of *two* things that they are identical is nonsense, and to say of *one* thing that it is identical with itself is to say nothing at all.

5.531 Ich schreibe also nicht „f(a,b).a = b“, sondern „f(a,a)“ (oder „f(b,b)“). Und nicht „f(a,b).~a = b“, sondern „f(a,b)“.

5.532 Und analog: Nicht „(∃x,y).f(x,y).x = y“, sondern „(∃x).f(x,x)“; und nicht „(∃x,y).f(x,y).~x = y“, sondern „(∃x,y).f(x,y)“.

(Also statt des Russellschen „(∃x,y).f(x,y)“

„(∃x,y).f(x,y).v.(∃x).f(x,x)“.)

5.5321 Statt „(x) : fx ⊃ x = a“ schreiben wir also z. B. „(∃x).fx.⊃.fa : ~ (∃x,y).fx.fy“.

Und der Satz: „Nur Ein x befriedigt f( )“, lautet: „(∃x).fx : ~ (∃x,y).fx.fy“.

5.533 Das Gleichheitszeichen ist also kein wesentlicher Bestandteil der Begriffsschrift.

5.534 Und nun sehen wir, daß Scheinsätze wie: „a = a“, „a = b.b = c.⊃a = c“, „(x).x = x“, „(∃x).x = a“, etc. sich in einer richtigen Begriffsschrift gar nicht hinschreiben lassen.

5.535 Damit erledigen sich auch alle Probleme, die an solche Scheinsätze geknüpft waren.

Alle Probleme, die Russells „Axiom of Infinity“ mit sich bringt, sind schon hier zu lösen.

Das, was das Axiom of Infinity sagen soll, würde sich in der Sprache dadurch ausdrücken, daß es unendlich viele Namen mit verschiedener Bedeutung gäbe.

5.5351 Es gibt gewisse Fälle, wo man in Versuchung gerät, Ausdrücke von der Form „a = a“ oder „p ⊃ p“ u. dgl. zu benützen. Und zwar geschieht dies, wenn man von dem Urbild: Satz, Ding, etc. reden möchte. So hat Russell in den „Principles of Mathematics“ den Unsinn „p ist ein Satz“ in Symbolen durch „p ⊃ p“ wiedergegeben und als Hypothese vor gewisse Sätze gestellt, damit deren Argumentstellen nur von Sätzen besetzt werden könnten.

(Es ist schon darum Unsinn, die Hypothese p ⊃ p vor einen Satz zu stellen, um ihm Argumente der richtigen Form zu sichern, weil die Hypothese für einen Nicht-Satz als Argument nicht falsch, sondern unsinnig wird, und

5.531 Thus I do not write '$f(a,b).a = b$', but '$f(a,a)$' (or '$f(b,b)$'); and not '$f(a,b).\sim a = b$', but '$f(a,b)$'.

5.532 And analogously I do not write '$(\exists x,y).f(x,y).x = y$', but '$(\exists x).f(x,x)$'; and not '$(\exists x,y).f(x,y).\sim x = y$', but '$(\exists x.y).f(x,y)$'.

(So Russell's '$(\exists x,y).fxy$' becomes

$$`(\exists x,y).f(x,y).\text{v}.(\exists x).f(x,x)\text{'.})$$

5.5321 Thus, for example, instead of '$(x):fx \supset x = a$' we write '$(\exists x).fx.\supset.fa : \sim(\exists x,y).fx.fy$'.

And the proposition, '*Only one* x satisfies $f(\ )$', will read '$(\exists x).fx : \sim(\exists x,y).fx.fy$'.

5.533 The identity-sign, therefore, is not an essential constituent of conceptual notation.

5.534 And now we see that in a correct conceptual notation pseudo-propositions like '$a = a$', '$a = b.b = c.\supset a = c$', '$(x).x = x$', '$(\exists x).x = a$', etc. cannot even be written down.

5.535 This also disposes of all the problems that were connected with such pseudo-propositions.

All the problems that Russell's 'axiom of infinity' brings with it can be solved at this point.

What the axiom of infinity is intended to say would express itself in language through the existence of infinitely many names with different meanings.

5.5351 There are certain cases in which one is tempted to use expressions of the form '$a = a$' or '$p \supset p$' and the like. In fact, this happens when one wants to talk about prototypes, e.g. about proposition, thing, etc. Thus in Russell's *Principles of Mathematics* 'p is a proposition'—which is nonsense—was given the symbolic rendering '$p \supset p$' and placed as an hypothesis in front of certain propositions in order to exclude from their argument-places everything but propositions.

(It is nonsense to place the hypothesis '$p \supset p$' in front of a proposition, in order to ensure that its arguments shall have the right form, if only because with a non-pro-

weil der Satz selbst durch die unrichtige Gattung von Argumenten unsinnig wird, also sich selbst ebenso gut, oder so schlecht, vor den unrechten Argumenten bewahrt wie die zu diesem Zweck angehängte sinnlose Hypothese.)

5.5352 Ebenso wollte man „Es gibt keine Dinge" ausdrücken durch „$\sim(\exists x) . x = x$". Aber selbst wenn dies ein Satz wäre — wäre er nicht auch wahr, wenn es zwar „Dinge gäbe", aber diese nicht mit sich selbst identisch wären?

5.54 In der allgemeinen Satzform kommt der Satz im Satze nur als Basis der Wahrheitsoperationen vor.

5.541 Auf den ersten Blick scheint es, als könne ein Satz in einem anderen auch auf andere Weise vorkommen.

Besonders in gewissen Satzformen der Psychologie, wie „A glaubt, daß p der Fall ist", oder „A denkt p", etc.

Hier scheint es nämlich oberflächlich, als stünde der Satz p zu einem Gegenstand A in einer Art von Relation.

(Und in der modernen Erkenntnistheorie (Russell, Moore, etc.) sind jene Sätze auch so aufgefaßt worden.)

5.542 Es ist aber klar, daß „A glaubt, daß p", „A denkt p", „A sagt p" von der Form „‚p' sagt p" sind: Und hier handelt es sich nicht um eine Zuordnung von einer Tatsache und einem Gegenstand, sondern um die Zuordnung von Tatsachen durch Zuordnung ihrer Gegenstände.

5.5421 Dies zeigt auch, daß die Seele — das Subjekt, etc. — wie sie in der heutigen oberflächlichen Psychologie aufgefaßt wird, ein Unding ist.

Eine zusammengesetzte Seele wäre nämlich keine Seele mehr.

5.5422 Die richtige Erklärung der Form des Satzes „A urteilt p" muß zeigen, daß es unmöglich ist, einen Unsinn zu urteilen. (Russells Theorie genügt dieser Bedingung nicht.)

position as argument the hypothesis becomes not false but nonsensical, and because arguments of the wrong kind make the proposition itself nonsensical, so that it preserves itself from wrong arguments just as well, or as badly, as the hypothesis without sense that was appended for that purpose.)

5.5352 In the same way people have wanted to express, 'There are no *things*', by writing '$\sim(\exists x).x = x$'. But even if this were a proposition, would it not be equally true if in fact 'there were things' but they were not identical with themselves?

5.54 In the general propositional form propositions occur in other propositions only as bases of truth-operations.

5.541 At first sight it looks as if it were also possible for one proposition to occur in another in a different way.

Particularly with certain forms of proposition in psychology, such as 'A believes that p is the case' and 'A has the thought p', etc.

For if these are considered superficially, it looks as if the proposition p stood in some kind of relation to an object A.

(And in modern theory of knowledge (Russell, Moore, etc.) these propositions have actually been construed in this way.)

5.542 It is clear, however, that 'A believes that p', 'A has the thought p', and 'A says p' are of the form '"p" says p': and this does not involve a correlation of a fact with an object, but rather the correlation of facts by means of the correlation of their objects.

5.5421 This shows too that there is no such thing as the soul—the subject, etc.—as it is conceived in the superficial psychology of the present day.

Indeed a composite soul would no longer be a soul.

5.5422 The correct explanation of the form of the proposition, 'A makes the judgement p', must show that it is impossible for a judgement to be a piece of nonsense. (Russell's theory does not satisfy this requirement.)

5.5423 Einen Komplex wahrnehmen heißt wahrnehmen, daß sich seine Bestandteile so und so zu einander verhalten.

Dies erklärt wohl auch, daß man die Figur

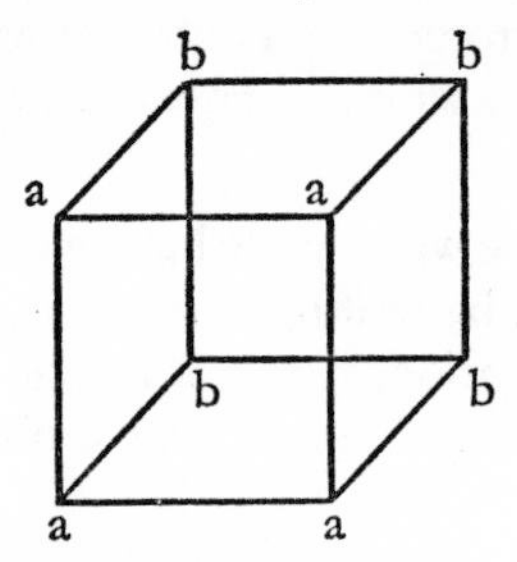

auf zweierlei Art als Würfel sehen kann; und alle ähnlichen Erscheinungen. Denn wir sehen eben wirklich zwei verschiedene Tatsachen.

(Sehe ich erst auf die Ecken a und nur flüchtig auf b, so erscheint a vorne; und umgekehrt.)

5.55 Wir müssen nun die Frage nach allen möglichen Formen der Elementarsätze a priori beantworten.

Der Elementarsatz besteht aus Namen. Da wir aber die Anzahl der Namen von verschiedener Bedeutung nicht angeben können, so können wir auch nicht die Zusammensetzung des Elementarsatzes angeben.

5.551 Unser Grundsatz ist, daß jede Frage, die sich überhaupt durch die Logik entscheiden läßt, sich ohne weiteres entscheiden lassen muß.

(Und wenn wir in die Lage kommen, ein solches Problem durch Ansehen der Welt beantworten zu müssen, so zeigt dies, daß wir auf grundfalscher Fährte sind.)

5.552 Die „Erfahrung", die wir zum Verstehen der Logik brauchen, ist nicht die, daß sich etwas so und so verhält, sondern, daß etwas *ist*: aber das ist eben *keine* Erfahrung.

Die Logik ist *vor* jeder Erfahrung — daß etwas *so* ist.

Sie ist vor dem Wie, nicht vor dem Was.

5.5521 Und wenn dies nicht so wäre, wie könnten wir die Logik anwenden? Man könnte sagen: Wenn es eine Logik gäbe, auch wenn es keine Welt gäbe, wie könnte es dann eine Logik geben, da es eine Welt gibt?

5.5423 To perceive a complex means to perceive that its constituents are related to one another in such and such a way.

This no doubt also explains why there are two possible ways of seeing the figure

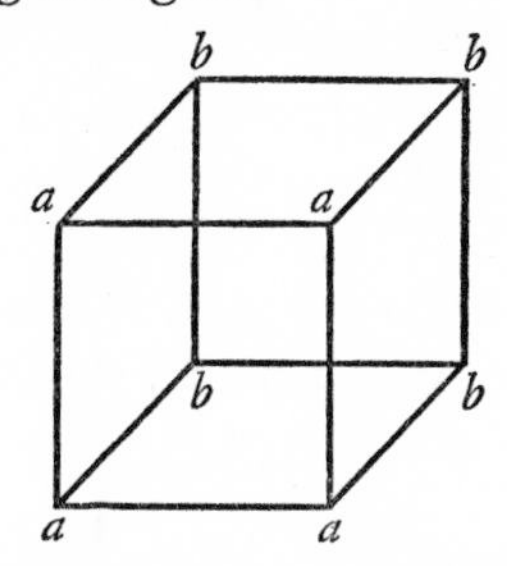

as a cube; and all similar phenomena. For we really see two different facts.

(If I look in the first place at the corners marked *a* and only glance at the *b*'s, then the *a*'s appear to be in front, and vice versa).

5.55 We now have to answer a priori the question about all the possible forms of elementary propositions.

Elementary propositions consist of names. Since, however, we are unable to give the number of names with different meanings, we are also unable to give the composition of elementary propositions.

5.551 Our fundamental principle is that whenever a question can be decided by logic at all it must be possible to decide it without more ado.

(And if we get into a position where we have to look at the world for an answer to such a problem, that shows that we are on a completely wrong track.)

5.552 The 'experience' that we need in order to understand logic is not that something or other is the state of things, but that something *is*: that, however, is *not* an experience.

Logic is *prior* to every experience—that something *is so*.

It is prior to the question 'How?', not prior to the question 'What?'

5.5521 And if this were not so, how could we apply logic? We might put it in this way: if there would be a logic even if there were no world, how then could there be a logic given that there is a world?

5.553 Russell sagte, es gäbe einfache Relationen zwischen verschiedenen Anzahlen von Dingen (Individuals). Aber zwischen welchen Anzahlen? Und wie soll sich das entscheiden?—Durch die Erfahrung?

(Eine ausgezeichnete Zahl gibt es nicht.)

5.554 Die Angabe jeder speziellen Form wäre vollkommen willkürlich.

5.5541 Es soll sich a priori angeben lassen, ob ich z. B. in die Lage kommen kann, etwas mit dem Zeichen einer 27-stelligen Relation bezeichnen zu müssen.

5.5542 Dürfen wir denn aber überhaupt so fragen? Können wir eine Zeichenform aufstellen und nicht wissen, ob ihr etwas entsprechen könne?

Hat die Frage einen Sinn: Was muß s e i n, damit etwas der-Fall-sein kann?

5.555 Es ist klar, wir haben vom Elementarsatz einen Begriff, abgesehen von seiner besonderen logischen Form.

Wo man aber Symbole nach einem System bilden kann, dort ist dieses System das logisch wichtige und nicht die einzelnen Symbole.

Und wie wäre es auch möglich, daß ich es in der Logik mit Formen zu tun hätte, die ich erfinden kann; sondern mit dem muß ich es zu tun haben, was es mir möglich macht, sie zu erfinden.

5.556 Eine Hierarchie der Formen der Elementarsätze kann es nicht geben. Nur was wir selbst konstruieren, können wir voraussehen.

5.5561 Die empirische Realität ist begrenzt durch die Gesamtheit der Gegenstände. Die Grenze zeigt sich wieder in der Gesamtheit der Elementarsätze.

Die Hierarchien sind, und müssen unabhängig von der Realität sein.

5.5562 Wissen wir aus rein logischen Gründen, daß es Elementarsätze geben muß, dann muß es jeder wissen, der die Sätze in ihrer unanalysierten Form versteht.

5.5563 Alle Sätze unserer Umgangssprache sind tatsächlich, so wie sie sind, logisch vollkommen geordnet.— Jenes

5.553 Russell said that there were simple relations between different numbers of things (individuals). But between what numbers? And how is this supposed to be decided? —By experience?

(There is no pre-eminent number.)

5.554 It would be completely arbitrary to give any specific form.

5.5541 It is supposed to be possible to answer a priori the question whether I can get into a position in which I need the sign for a 27-termed relation in order to signify something.

5.5542 But is it really legitimate even to ask such a question? Can we set up a form of sign without knowing whether anything can correspond to it?

Does it make sense to ask what there must *be* in order that something can be the case?

5.555 Clearly we have some concept of elementary propositions quite apart from their particular logical forms.

But when there is a system by which we can create symbols, the system is what is important for logic and not the individual symbols.

And anyway, is it really possible that in logic I should have to deal with forms that I can invent? What I have to deal with must be that which makes it possible for me to invent them.

5.556 There cannot be a hierarchy of the forms of elementary propositions. We can foresee only what we ourselves construct.

5.5561 Empirical reality is limited by the totality of objects. The limit also makes itself manifest in the totality of elementary propositions.

Hierarchies are and must be independent of reality.

5.5562 If we know on purely logical grounds that there must be elementary propositions, then everyone who understands propositions in their unanalysed form must know it.

5.5563 In fact, all the propositions of our everyday language, just as they stand, are in perfect logical order.—That

Einfachste, was wir hier angeben sollen, ist nicht ein Gleichnis der Wahrheit, sondern die volle Wahrheit selbst.

(Unsere Probleme sind nicht abstrakt, sondern vielleicht die konkretesten, die es gibt.)

5.557 Die *Anwendung* der Logik entscheidet darüber, welche Elementarsätze es gibt.

Was in der Anwendung liegt, kann die Logik nicht vorausnehmen.

Das ist klar: Die Logik darf mit ihrer Anwendung nicht kollidieren.

Aber die Logik muß sich mit ihrer Anwendung berühren.

Also dürfen die Logik und ihre Anwendung einander nicht übergreifen.

5.5571 Wenn ich die Elementarsätze nicht a priori angeben kann, dann muß es zu offenbarem Unsinn führen, sie angeben zu wollen.

5.6 *Die Grenzen meiner Sprache* bedeuten die Grenzen meiner Welt.

5.61 Die Logik erfüllt die Welt; die Grenzen der Welt sind auch ihre Grenzen.

Wir können also in der Logik nicht sagen: Das und das gibt es in der Welt, jenes nicht.

Das würde nämlich scheinbar voraussetzen, daß wir gewisse Möglichkeiten ausschließen, und dies kann nicht der Fall sein, da sonst die Logik über die Grenzen der Welt hinaus müßte; wenn sie nämlich diese Grenzen auch von der anderen Seite betrachten könnte.

Was wir nicht denken können, das können wir nicht denken; wir können also auch nicht *sagen*, was wir nicht denken können.

5.62 Diese Bemerkung gibt den Schlüssel zur Entscheidung der Frage, inwieweit der Solipsismus eine Wahrheit ist.

Was der Solipsismus nämlich *meint*, ist ganz richtig, nur läßt es sich nicht *sagen*, sondern es zeigt sich.

Daß die Welt *meine* Welt ist, das zeigt sich darin, daß die Grenzen *der* Sprache (der Sprache, die allein ich verstehe) die Grenzen *meiner* Welt bedeuten.

5.621 Die Welt und das Leben sind Eins.

utterly simple thing, which we have to formulate here, is not a likeness of the truth, but the truth itself in its entirety.

(Our problems are not abstract, but perhaps the most concrete that there are.)

5.557 The *application* of logic decides what elementary propositions there are.

What belongs to its application, logic cannot anticipate.

It is clear that logic must not clash with its application.

But logic has to be in contact with its application.

Therefore logic and its application must not overlap.

5.5571 If I cannot say a priori what elementary propositions there are, then the attempt to do so must lead to obvious nonsense.

5.6 *The limits of my language* mean the limits of my world.

5.61 Logic pervades the world: the limits of the world are also its limits.

So we cannot say in logic, 'The world has this in it, and this, but not that.'

For that would appear to presuppose that we were excluding certain possibilities, and this cannot be the case, since it would require that logic should go beyond the limits of the world; for only in that way could it view those limits from the other side as well.

We cannot think what we cannot think; so what we cannot think we cannot *say* either.

5.62 This remark provides the key to the problem, how much truth there is in solipsism.

For what the solipsist *means* is quite correct; only it cannot be *said*, but makes itself manifest.

The world is *my* world: this is manifest in the fact that the limits of *language* (of that language which alone I understand) mean the limits of *my* world.

5.621 The world and life are one.

5.63 Ich bin meine Welt. (Der Mikrokosmos.)

5.631 Das denkende, vorstellende, Subjekt gibt es nicht.

Wenn ich ein Buch schriebe „Die Welt, wie ich sie vorfand", so wäre darin auch über meinen Leib zu berichten und zu sagen, welche Glieder meinem Willen unterstehen und welche nicht, etc., dies ist nämlich eine Methode, das Subjekt zu isolieren, oder vielmehr zu zeigen, daß es in einem wichtigen Sinne kein Subjekt gibt: Von ihm allein nämlich könnte in diesem Buche *nicht* die Rede sein.—

5.632 Das Subjekt gehört nicht zur Welt, sondern es ist eine Grenze der Welt.

5.633 Wo *in* der Welt ist ein metaphysisches Subjekt zu merken?

Du sagst, es verhält sich hier ganz wie mit Auge und Gesichtsfeld. Aber das Auge siehst du wirklich *nicht*.

Und nichts *am Gesichtsfeld* läßt darauf schließen, daß es von einem Auge gesehen wird.

5.6331 Das Gesichtsfeld hat nämlich nicht etwa eine solche Form:

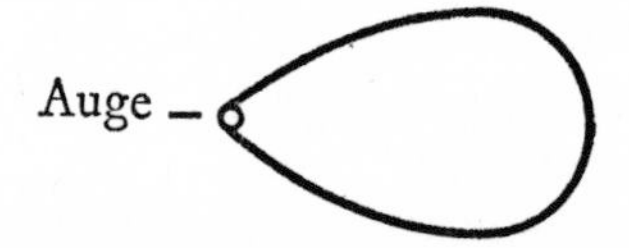

5.634 Das hängt damit zusammen, daß kein Teil unserer Erfahrung auch a priori ist.

Alles, was wir sehen, könnte auch anders sein.

Alles, was wir überhaupt beschreiben können, könnte auch anders sein.

Es gibt keine Ordnung der Dinge a priori.

5.64 Hier sieht man, daß der Solipsismus, streng durchgeführt, mit dem reinen Realismus zusammenfällt. Das Ich des Solipsismus schrumpft zum ausdehnungslosen Punkt zusammen, und es bleibt die ihm koordinierte Realität.

5.641 Es gibt also wirklich einen Sinn, in welchem in der Philosophie nichtpsychologisch vom Ich die Rede sein kann.

Das Ich tritt in die Philosophie dadurch ein, daß „die Welt meine Welt ist".

5.63 I am my world. (The microcosm.)

5.631 There is no such thing as the subject that thinks or entertains ideas.

If I wrote a book called *The World as I found it*, I should have to include a report on my body, and should have to say which parts were subordinate to my will, and which were not, etc., this being a method of isolating the subject, or rather of showing that in an important sense there is no subject; for it alone could *not* be mentioned in that book.—

5.632 The subject does not belong to the world: rather, it is a limit of the world.

5.633 Where *in* the world is a metaphysical subject to be found?

You will say that this is exactly like the case of the eye and the visual field. But really you do *not* see the eye.

And nothing *in the visual field* allows you to infer that it is seen by an eye.

5.6331 For the form of the visual field is surely not like this

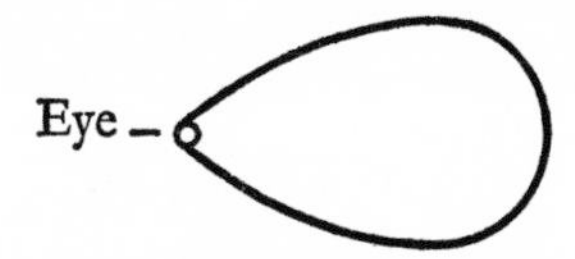

5.634 This is connected with the fact that no part of our experience is at the same time a priori.

Whatever we see could be other than it is.

Whatever we can describe at all could be other than it is.

There is no a priori order of things.

5.64 Here it can be seen that solipsism, when its implications are followed out strictly, coincides with pure realism. The self of solipsism shrinks to a point without extension, and there remains the reality co-ordinated with it.

5.641 Thus there really is a sense in which philosophy can talk about the self in a non-psychological way.

What brings the self into philosophy is the fact that 'the world is my world'.

Das philosophische Ich ist nicht der Mensch, nicht der menschliche Körper, oder die menschliche Seele, von der die Psychologie handelt, sondern das metaphysische Subjekt, die Grenze — nicht ein Teil — der Welt.

6 Die allgemeine Form der Wahrheitsfunktion ist: $[\bar{p}, \bar{\xi}, N(\bar{\xi})]$.

Dies ist die allgemeine Form des Satzes.

6.001 Dies sagt nichts anderes als, daß jeder Satz ein Resultat der successiven Anwendung der Operation $N(\bar{\xi})$ auf die Elementarsätze ist.

6.002 Ist die allgemeine Form gegeben, wie ein Satz gebaut ist, so ist damit auch schon die allgemeine Form davon gegeben, wie aus einem Satz durch eine Operation ein anderer erzeugt werden kann.

6.01 Die allgemeine Form der Operation $\Omega'(\bar{\eta})$ ist also:

$$[\bar{\xi}, N(\bar{\xi})]'(\bar{\eta}) \quad (= [\bar{\eta}, \bar{\xi}, N(\bar{\xi})]).$$

Das ist die allgemeinste Form des Überganges von einem Satz zum anderen.

6.02 Und **so** kommen wir zu den Zahlen: Ich definiere

$$x = \Omega^{0}{}'x \text{ Def.}$$

und

$$\Omega'\Omega^{\nu}{}'x = \Omega^{\nu+1}{}'x \text{ Def.}$$

Nach diesen Zeichenregeln schreiben wir also die Reihe

$$x, \ \Omega'x, \ \Omega'\Omega'x, \ \Omega'\Omega'\Omega'x, \ldots,$$

so

$$\Omega^{0}{}'x, \ \Omega^{0+1}{}'x, \ \Omega^{0+1+1}{}'x, \ \Omega^{0+1+1+1}{}'x, \ldots.$$

Also schreibe ich — statt „$[x, \xi, \Omega'\xi]$"—

„$[\Omega^{0}{}'x, \Omega^{\nu}{}'x, \Omega^{\nu+1}{}'x]$".

Und definiere

$$0+1 = 1 \text{ Def.},$$
$$0+1+1 = 2 \text{ Def.},$$
$$0+1+1+1 = 3 \text{ Def.},$$
(u.s.f.)

The philosophical self is not the human being, not the human body, or the human soul, with which psychology deals, but rather the metaphysical subject, the limit of the world—not a part of it.

6 The general form of a truth-function is $[\bar{p}, \bar{\xi}, N(\bar{\xi})]$. This is the general form of a proposition.

6.001 What this says is just that every proposition is a result of successive applications to elementary propositions of the operation $N(\bar{\xi})$.

6.002 If we are given the general form according to which propositions are constructed, then with it we are also given the general form according to which one proposition can be generated out of another by means of an operation.

6.01 Therefore the general form of an operation $\Omega'(\bar{\eta})$ is

$$[\bar{\xi}, N(\bar{\xi})]'(\bar{\eta}) \quad (= [\bar{\eta}, \bar{\xi}, N(\bar{\xi})]).$$

This is the most general form of transition from one proposition to another.

6.02 And *this* is how we arrive at numbers. I give the following definitions

$$x = \Omega^{0}{}'x \text{ Def.},$$

$$\Omega'\Omega^{\nu}{}'x = \Omega^{\nu+1}{}'x \text{ Def.}$$

So, in accordance with these rules, which deal with signs, we write the series

$$x, \ \Omega'x, \ \Omega'\Omega'x, \ \Omega'\Omega'\Omega'x, \ldots,$$

in the following way

$$\Omega^{0}{}'x, \ \Omega^{0+1}{}'x, \ \Omega^{0+1+1}{}'x, \ \Omega^{0+1+1+1}{}'x, \ldots.$$

Therefore, instead of '$[x, \xi, \Omega'\xi]$',

I write '$[\Omega^{0}{}'x, \Omega^{\nu}{}'x, \Omega^{\nu+1}{}'x]$'.

And I give the following definitions

$$0+1 = 1 \text{ Def.},$$
$$0+1+1 = 2 \text{ Def.},$$
$$0+1+1+1 = 3 \text{ Def.},$$
(and so on).

6.021 Die Zahl ist der Exponent einer Operation.

6.022 Der Zahlbegriff ist nichts anderes als das Gemeinsame aller Zahlen, die allgemeine Form der Zahl.

Der Zahlbegriff ist die variable Zahl.

Und der Begriff der Zahlengleichheit ist die allgemeine Form aller speziellen Zahlengleichheiten.

6.03 Die allgemeine Form der ganzen Zahl ist: $[0, \xi, \xi+1]$.

6.031 Die Theorie der Klassen ist in der Mathematik ganz überflüssig.

Dies hängt damit zusammen, daß die Allgemeinheit, welche wir in der Mathematik brauchen, nicht die *zufällige* ist.

6.1 Die Sätze der Logik sind Tautologien.

6.11 Die Sätze der Logik sagen also nichts. (Sie sind die analytischen Sätze.)

6.111 Theorien, die einen Satz der Logik gehaltvoll erscheinen lassen, sind immer falsch. Man könnte z. B. glauben, daß die Worte „wahr" und „falsch" zwei Eigenschaften unter anderen Eigenschaften bezeichnen, und da erschiene es als eine merkwürdige Tatsache, daß jeder Satz eine dieser Eigenschaften besitzt. Das scheint nun nichts weniger als selbstverständlich zu sein, ebensowenig selbstverständlich, wie etwa der Satz: „Alle Rosen sind entweder gelb oder rot", klänge, auch wenn er wahr wäre. Ja, jener Satz bekommt nun ganz den Charakter eines naturwissenschaftlichen Satzes und dies ist das sichere Anzeichen dafür, daß er falsch aufgefaßt wurde.

6.112 Die richtige Erklärung der logischen Sätze muß ihnen eine einzigartige Stellung unter allen Sätzen geben.

6.113 Es ist das besondere Merkmal der logischen Sätze, daß man am Symbol allein erkennen kann, daß sie wahr sind, und diese Tatsache schließt die ganze Philosophie der Logik in sich. Und so ist es auch eine der wichtigsten Tatsachen, daß sich die Wahrheit oder Falschheit der nichtlogischen Sätze *nicht* am Satz allein erkennen läßt.

6.12 Daß die Sätze der Logik Tautologien sind, das *zeigt* die formalen — logischen — Eigenschaften der Sprache, der Welt.

6.021 A number is the exponent of an operation.

6.022 The concept of number is simply what is common to all numbers, the general form of a number.

The concept of number is the variable number.

And the concept of numerical equality is the general form of all particular cases of numerical equality.

6.03 The general form of an integer is $[0, \xi, \xi+1]$.

6.031 The theory of classes is completely superfluous in mathematics.

This is connected with the fact that the generality required in mathematics is not *accidental* generality.

6.1 The propositions of logic are tautologies.

6.11 Therefore the propositions of logic say nothing. (They are the analytic propositions.)

6.111 All theories that make a proposition of logic appear to have content are false. One might think, for example, that the words 'true' and 'false' signified two properties among other properties, and then it would seem to be a remarkable fact that every proposition possessed one of these properties. On this theory it seems to be anything but obvious, just as, for instance, the proposition, 'All roses are either yellow or red', would not sound obvious even if it were true. Indeed, the logical proposition acquires all the characteristics of a proposition of natural science and this is the sure sign that it has been construed wrongly.

6.112 The correct explanation of the propositions of logic must assign to them a unique status among all propositions.

6.113 It is the peculiar mark of logical propositions that one can recognize that they are true from the symbol alone, and this fact contains in itself the whole philosophy of logic. And so too it is a very important fact that the truth or falsity of non-logical propositions *cannot* be recognized from the propositions alone.

6.12 The fact that the propositions of logic are tautologies *shows* the formal—logical—properties of language and the world.

Daß ihre Bestandteile s o verknüpft eine Tautologie ergeben, das charakterisiert die Logik ihrer Bestandteile.

Damit Sätze, auf bestimmte Art und Weise verknüpft, eine Tautologie ergeben, dazu müssen sie bestimmte Eigenschaften der Struktur haben. Daß sie s o verbunden eine Tautologie ergeben, zeigt also, daß sie diese Eigenschaften der Struktur besitzen.

6.1201 Daß z. B. die Sätze „p" und „~p" in der Verbindung „~(p.~p)" eine Tautologie ergeben, zeigt, daß sie einander widersprechen. Daß die Sätze „p ⊃ q", „p" und „q" in der Form „(p ⊃ q).(p):⊃:(q)" miteinander verbunden eine Tautologie ergeben, zeigt, daß q aus p und p ⊃ q folgt. Daß „(x).fx:⊃:fa" eine Tautologie ist, daß fa aus (x).fx folgt. Etc. etc.

6.1202 Es ist klar, daß man zu demselben Zweck statt der Tautologien auch die Kontradiktionen verwenden könnte.

6.1203 Um eine Tautologie als solche zu erkennen, kann man sich, in den Fällen, in welchen in der Tautologie keine Allgemeinheitsbezeichnung vorkommt, folgender anschaulichen Methode bedienen: Ich schreibe statt „p" „q", „r", etc. „WpF", „WqF", „WrF", etc. Die Wahrheitskombinationen drücke ich durch Klammern aus, z. B.:

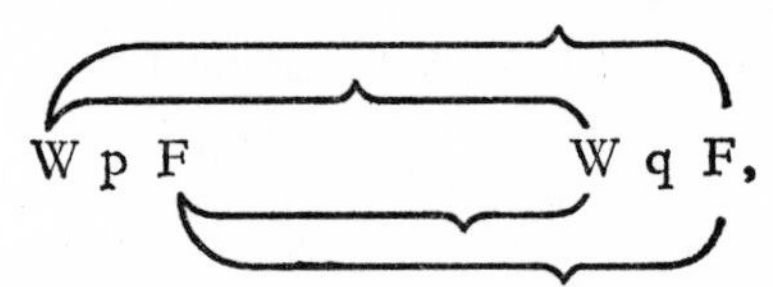

und die Zuordnung der Wahr- oder Falschheit des ganzen Satzes und der Wahrheitskombinationen der Wahrheitsargumente durch Striche auf folgende Weise:

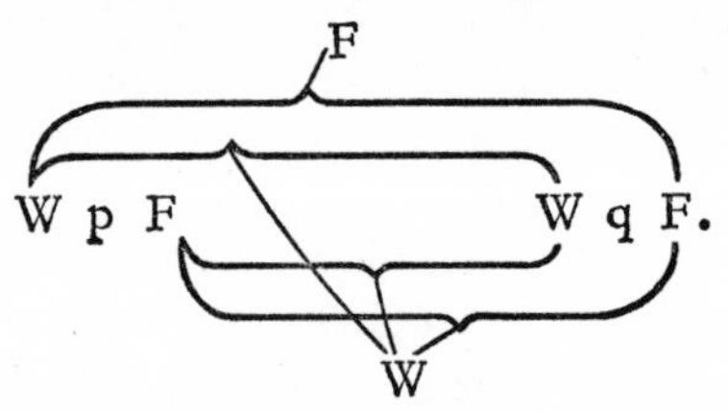

Dies Zeichen würde also z. B. den Satz p ⊃ q darstellen. Nun will ich z. B. den Satz ~(p.~p) (Gesetz des Wider-

The fact that a tautology is yielded by *this particular way* of connecting its constituents characterizes the logic of its constituents.

If propositions are to yield a tautology when they are connected in a certain way, they must have certain structural properties. So their yielding a tautology when combined *in this way* shows that they possess these structural properties.

6.1201 For example, the fact that the propositions 'p' and '$\sim p$' in the combination '$\sim(p \,.\, \sim p)$' yield a tautology shows that they contradict one another. The fact that the propositions '$p \supset q$', 'p', and 'q', combined with one another in the form '$(p \supset q) \,.\, (p) : \supset : (q)$', yield a tautology shows that q follows from p and $p \supset q$. The fact that '$(x) \,.\, fx : \supset : fa$' is a tautology shows that fa follows from $(x) \,.\, fx$. Etc. etc.

6.1202 It is clear that one could achieve the same purpose by using contradictions instead of tautologies.

6.1203 In order to recognize an expression as a tautology, in cases where no generality-sign occurs in it, one can employ the following intuitive method: instead of 'p', 'q', 'r', etc. I write 'TpF', 'TqF', 'TrF', etc. Truth-combinations I express by means of brackets, e.g.

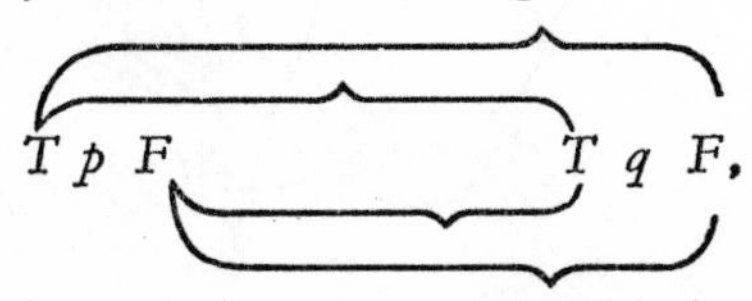

and I use lines to express the correlation of the truth or falsity of the whole proposition with the truth-combinations of its truth-arguments, in the following way

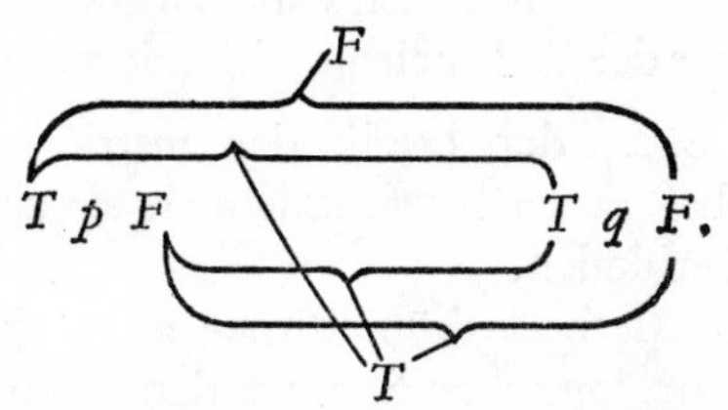

So this sign, for instance, would represent the proposition $p \supset q$. Now, by way of example, I wish to examine the proposition $\sim(p \,.\, \sim p)$ (the law of contradiction) in order

spruchs) daraufhin untersuchen, ob er eine Tautologie ist. Die Form „$\sim\xi$" wird in unserer Notation

W

„WξF"

F

geschrieben; die Form „$\xi . \eta$" so:

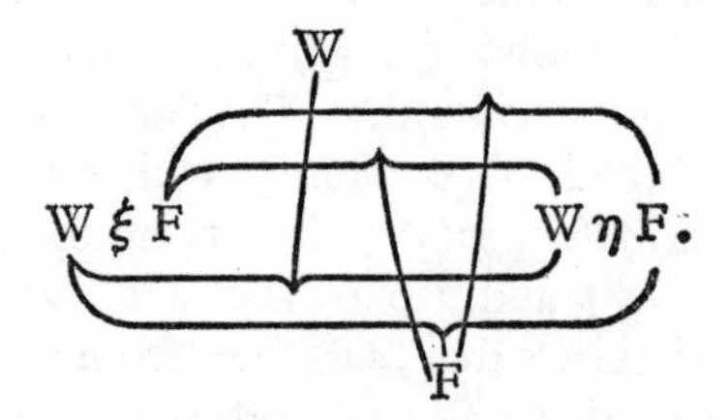

Daher lautet der Satz $\sim(p . \sim q)$ so:

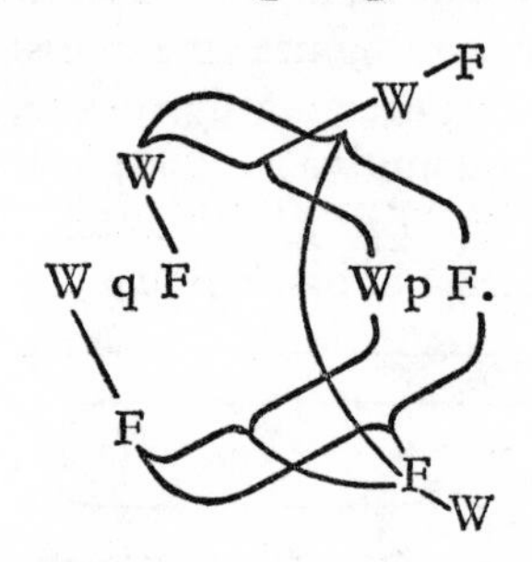

Setzen wir hier statt „q" „p" ein und untersuchen die Verbindung der äußersten W und F mit den innersten, so ergibt sich, daß die Wahrheit des ganzen Satzes allen Wahrheitskombinationen seines Argumentes, seine Falschheit keiner der Wahrheitskombinationen zugeordnet ist.

6.121 Die Sätze der Logik demonstrieren die logischen Eigenschaften der Sätze, indem sie sie zu nichtssagenden Sätzen verbinden.

Diese Methode könnte man auch eine Nullmethode nennen. Im logischen Satz werden Sätze miteinander ins Gleichgewicht gebracht und der Zustand des Gleichgewichts zeigt dann an, wie diese Sätze logisch beschaffen sein müssen.

to determine whether it is a tautology. In our notation the form '$\sim\xi$' is written as

$$\begin{matrix} T \\ \diagdown \\ \text{'}T\xi F\text{'}, \\ \diagdown \\ F \end{matrix}$$

and the form '$\xi . \eta$' as

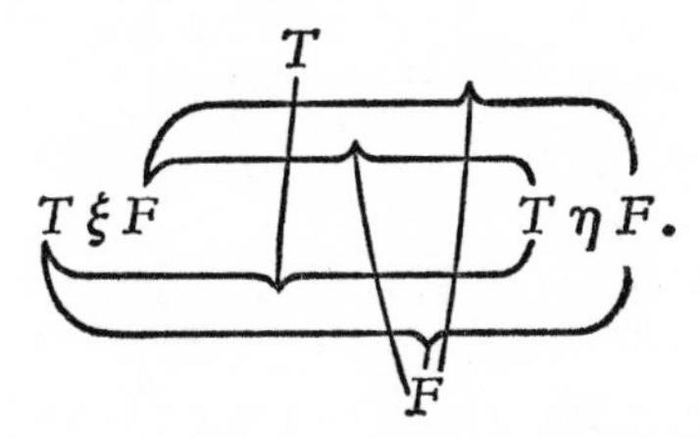

Hence the proposition $\sim(p . \sim q)$ reads as follows

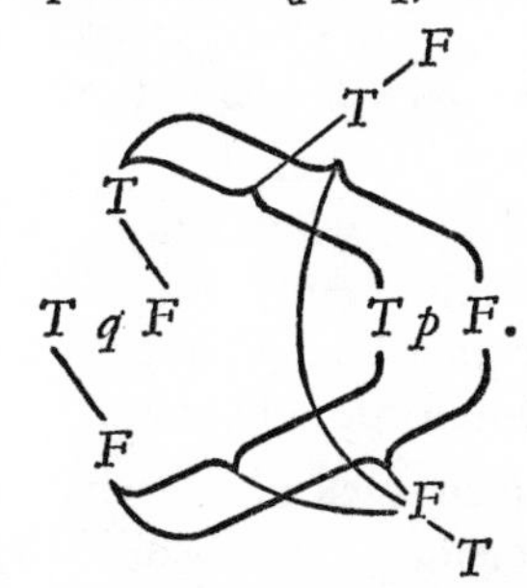

If we here substitute 'p' for 'q' and examine how the outermost T and F are connected with the innermost ones, the result will be that the truth of the whole proposition is correlated with *all* the truth-combinations of its argument, and its falsity with none of the truth-combinations.

6.121 The propositions of logic demonstrate the logical properties of propositions by combining them so as to form propositions that say nothing.

This method could also be called a zero-method. In a logical proposition, propositions are brought into equilibrium with one another, and the state of equilibrium then indicates what the logical constitution of these propositions must be.

6.122 Daraus ergibt sich, daß wir auch ohne die logischen Sätze auskommen können, da wir ja in einer entsprechenden Notation die formalen Eigenschaften der Sätze durch das bloße Ansehen dieser Sätze erkennen können.

6.1221 Ergeben z. B. zwei Sätze „p“ und „q“ in der Verbindung „$p \supset q$“ eine Tautologie, so ist klar, daß q aus p folgt.

Daß z. B. „q“ aus „$p \supset q . p$“ folgt, ersehen wir aus diesen beiden Sätzen selbst, aber wir können es auch *so* zeigen, indem wir sie zu „$p \supset q . p : \supset : q$“ verbinden und nun zeigen, daß dies eine Tautologie ist.

6.1222 Dies wirft ein Licht auf die Frage, warum die logischen Sätze nicht durch die Erfahrung bestätigt werden können, ebensowenig wie sie durch die Erfahrung widerlegt werden können. Nicht nur muß ein Satz der Logik durch keine mögliche Erfahrung widerlegt werden können, sondern er darf auch nicht durch eine solche bestätigt werden können.

6.1223 Nun wird klar, warum man oft fühlte, als wären die „logischen Wahrheiten“ von uns zu „*fordern*“: Wir können sie nämlich insofern fordern, als wir eine genügende Notation fordern können.

6.1224 Es wird jetzt auch klar, warum die Logik die Lehre von den Formen und vom Schließen genannt wurde.

6.123 Es ist klar: Die logischen Gesetze dürfen nicht selbst wieder logischen Gesetzen unterstehen.

(Es gibt nicht, wie Russell meinte, für jede „Type“ ein eigenes Gesetz des Widerspruches, sondern Eines genügt, da es auf sich selbst nicht angewendet wird.)

6.1231 Das Anzeichen des logischen Satzes ist *nicht* die Allgemeingültigkeit.

Allgemein sein heißt ja nur: zufälligerweise für alle Dinge gelten. Ein unverallgemeinerter Satz kann ja ebensowohl tautologisch sein als ein verallgemeinerter.

6.1232 Die logische Allgemeingültigkeit könnte man wesentlich nennen, im Gegensatz zu jener zufälligen, etwa des Satzes: „Alle Menschen sind sterblich“. Sätze wie Russells

6.122 It follows from this that we can actually do without logical propositions; for in a suitable notation we can in fact recognize the formal properties of propositions by mere inspection of the propositions themselves.

6.1221 If, for example, two propositions 'p' and 'q' in the combination '$p \supset q$' yield a tautology, then it is clear that q follows from p.

For example, we see from the two propositions themselves that 'q' follows from '$p \supset q . p$', but it is also possible to show it in *this* way: we combine them to form '$p \supset q . p : \supset : q$', and then show that this is a tautology.

6.1222 This throws some light on the question why logical propositions cannot be confirmed by experience any more than they can be refuted by it. Not only must a proposition of logic be irrefutable by any possible experience, but it must also be unconfirmable by any possible experience.

6.1223 Now it becomes clear why people have often felt as if it were for us to '*postulate*' the 'truths of logic'. The reason is that we can postulate them in so far as we can postulate an adequate notation.

6.1224 It also becomes clear now why logic was called the theory of forms and of inference.

6.123 Clearly the laws of logic cannot in their turn be subject to laws of logic.

(There is not, as Russell thought, a special law of contradiction for each 'type'; one law is enough, since it is not applied to itself.)

6.1231 The mark of a logical proposition is *not* general validity.

To be general means no more than to be accidentally valid for all things. An ungeneralized proposition can be tautological just as well as a generalized one.

6.1232 The general validity of logic might be called essential, in contrast with the accidental general validity of such propositions as 'All men are mortal'. Propositions like

„Axiom of Reducibility“ sind nicht logische Sätze, und dies erklärt unser Gefühl: Daß sie, wenn wahr, so doch nur durch einen günstigen Zufall wahr sein könnten.

6.1233 Es läßt sich eine Welt denken, in der das Axiom of Reducibility nicht gilt. Es ist aber klar daß die Logik nichts mit der Frage zu schaffen hat, ob unsere Welt wirklich so ist oder nicht.

6.124 Die logischen Sätze beschreiben das Gerüst der Welt, oder vielmehr, sie stellen es dar. Sie „handeln“ von nichts. Sie setzen voraus, daß Namen Bedeutung, und Elementarsätze Sinn haben: Und dies ist ihre Verbindung mit der Welt. Es ist klar, daß es etwas über die Welt anzeigen muß, daß gewisse Verbindungen von Symbolen — welche wesentlich einen bestimmten Charakter haben — Tautologien sind. Hierin liegt das Entscheidende. Wir sagten, manches an den Symbolen, die wir gebrauchen, wäre willkürlich, manches nicht. In der Logik drückt nur dieses aus: Das heißt aber, in der Logik drücken nicht *wir* mit Hilfe der Zeichen aus, was wir wollen, sondern in der Logik sagt die Natur der naturnotwendigen Zeichen selbst aus: Wenn wir die logische Syntax irgend einer Zeichensprache kennen, dann sind bereits alle Sätze der Logik gegeben.

6.125 Es ist möglich, und zwar auch nach der alten Auffassung der Logik, von vornherein eine Beschreibung aller „wahren“ logischen Sätze zu geben.

6.1251 Darum kann es in der Logik auch *nie* Überraschungen geben.

6.126 Ob ein Satz der Logik angehört, kann man berechnen, indem man die logischen Eigenschaften des *Symbols* berechnet.

Und dies tun wir, wenn wir einen logischen Satz „beweisen“. Denn, ohne uns um einen Sinn und eine Bedeutung zu kümmern, bilden wir den logischen Satz aus anderen nach bloßen *Zeichenregeln*.

Der Beweis der logischen Sätze besteht darin, daß wir sie aus anderen logischen Sätzen durch successive Anwendung gewisser Operationen entstehen lassen, die aus den

Russell's 'axiom of reducibility' are not logical propositions, and this explains our feeling that, even if they were true, their truth could only be the result of a fortunate accident.

6.1233 It is possible to imagine a world in which the axiom of reducibility is not valid. It is clear, however, that logic has nothing to do with the question whether our world really is like that or not.

6.124 The propositions of logic describe the scaffolding of the world, or rather they represent it. They have no 'subject-matter'. They presuppose that names have meaning and elementary propositions sense; and that is their connexion with the world. It is clear that something about the world must be indicated by the fact that certain combinations of symbols—whose essence involves the possession of a determinate character—are tautologies. This contains the decisive point. We have said that some things are arbitrary in the symbols that we use and that some things are not. In logic it is only the latter that express: but that means that logic is not a field in which *we* express what we wish with the help of signs, but rather one in which the nature of the absolutely necessary signs speaks for itself. If we know the logical syntax of any sign-language, then we have already been given all the propositions of logic.

6.125 It is possible—indeed possible even according to the old conception of logic—to give in advance a description of all 'true' logical propositions.

6.1251 Hence there can *never* be surprises in logic.

6.126 One can calculate whether a proposition belongs to logic, by calculating the logical properties of the *symbol*.

And this is what we do when we 'prove' a logical proposition. For, without bothering about sense or meaning, we construct the logical proposition out of others using only *rules that deal with signs*.

The proof of logical propositions consists in the following process: we produce them out of other logical propositions by successively applying certain operations that always generate further tautologies out of the initial

ersten immer wieder Tautologien erzeugen. (Und zwar f o l g e n aus einer Tautologie nur Tautologien.)

Natürlich ist diese Art zu zeigen, daß ihre Sätze Tautologien sind, der Logik durchaus unwesentlich. Schon darum, weil die Sätze, von welchen der Beweis ausgeht, ja ohne Beweis zeigen müssen, daß sie Tautologien sind.

6.1261 In der Logik sind Prozeß und Resultat äquivalent. (Darum keine Überraschung.)

6.1262 Der Beweis in der Logik ist nur ein mechanisches Hilfsmittel zum leichteren Erkennen der Tautologie, wo sie kompliziert ist.

6.1263 Es wäre ja auch zu merkwürdig, wenn man einen sinnvollen Satz l o g i s c h aus anderen beweisen könnte, und einen logischen Satz a u c h. Es ist von vornherein klar, daß der logische Beweis eines sinnvollen Satzes und der Beweis i n der Logik zwei ganz verschiedene Dinge sein müssen.

6.1264 Der sinnvolle Satz sagt etwas aus, und sein Beweis zeigt, daß es so ist; in der Logik ist jeder Satz die Form eines Beweises.

Jeder Satz der Logik ist ein in Zeichen dargestellter modus ponens. (Und den modus ponens kann man nicht durch einen Satz ausdrücken.)

6.1265 Immer kann man die Logik so auffassen, daß jeder Satz sein eigener Beweis ist.

6.127 Alle Sätze der Logik sind gleichberechtigt, es gibt unter ihnen nicht wesentlich Grundgesetze und abgeleitete Sätze.

Jede Tautologie zeigt selbst, daß sie eine Tautologie ist.

6.1271 Es ist klar, daß die Anzahl der „logischen Grundgesetze“ willkürlich ist, denn man könnte die Logik ja aus Einem Grundgesetz ableiten, indem man einfach z. B. aus Freges Grundgesetzen das logische Produkt bildet. (Frege würde vielleicht sagen, daß dieses Grundgesetz nun nicht mehr unmittelbar einleuchte. Aber es ist merkwürdig, daß ein so exakter Denker wie Frege sich auf den Grad des Einleuchtens als Kriterium des logischen Satzes berufen hat.)

ones. (And in fact only tautologies *follow* from a tautology.)

Of course this way of showing that the propositions of logic are tautologies is not at all essential to logic, if only because the propositions from which the proof starts must show without any proof that they are tautologies.

6.1261 In logic process and result are equivalent. (Hence the absence of surprise.)

6.1262 Proof in logic is merely a mechanical expedient to facilitate the recognition of tautologies in complicated cases.

6.1263 Indeed, it would be altogether too remarkable if a proposition that had sense could be proved *logically* from others, and *so too* could a logical proposition. It is clear from the start that a logical proof of a proposition that has sense and a proof *in* logic must be two entirely different things.

6.1264 A proposition that has sense states something, which is shown by its proof to be so. In logic every proposition is the form of a proof.

Every proposition of logic is a modus ponens represented in signs. (And one cannot express the modus ponens by means of a proposition.)

6.1265 It is always possible to construe logic in such a way that every proposition is its own proof.

6.127 All the propositions of logic are of equal status: it is not the case that some of them are essentially primitive propositions and others essentially derived propositions.

Every tautology itself shows that it is a tautology.

6.1271 It is clear that the number of the 'primitive propositions of logic' is arbitrary, since one could derive logic from a single primitive proposition, e.g. by simply constructing the logical product of Frege's primitive propositions. (Frege would perhaps say that we should then no longer have an immediately self-evident primitive proposition. But it is remarkable that a thinker as rigorous as Frege appealed to the degree of self-evidence as the criterion of a logical proposition.)

6.13 Die Logik ist keine Lehre, sondern ein Spiegelbild der Welt.

Die Logik ist transzendental.

6.2 Die Mathematik ist eine logische Methode.

Die Sätze der Mathematik sind Gleichungen, also Scheinsätze.

6.21 Der Satz der Mathematik drückt keinen Gedanken aus.

6.211 Im Leben ist es ja nie der mathematische Satz, den wir brauchen, sondern wir benützen den mathematischen Satz n u r, um aus Sätzen, welche nicht der Mathematik angehören, auf andere zu schließen, welche gleichfalls nicht der Mathematik angehören.

(In der Philosophie führt die Frage: „Wozu gebrauchen wir eigentlich jenes Wort, jenen Satz?" immer wieder zu wertvollen Einsichten.)

6.22 Die Logik der Welt, die die Sätze der Logik in den Tautologien zeigen, zeigt die Mathematik in den Gleichungen.

6.23 Wenn zwei Ausdrücke durch das Gleichheitszeichen verbunden werden, so heißt das, sie sind durch einander ersetzbar. Ob dies aber der Fall ist, muß sich an den beiden Ausdrücken selbst zeigen.

Es charakterisiert die logische Form zweier Ausdrücke, daß sie durch einander ersetzbar sind.

6.231 Es ist eine Eigenschaft der Bejahung, daß man sie als doppelte Verneinung auffassen kann.

Es ist eine Eigenschaft von „1+1+1+1", daß man es als „(1+1)+(1+1)" auffassen kann.

6.232 Frege sagt, die beiden Ausdrücke haben dieselbe Bedeutung, aber verschiedenen Sinn.

Das Wesentliche an der Gleichung ist aber, daß sie nicht notwendig ist, um zu zeigen, daß die beiden Ausdrücke, die das Gleichheitszeichen verbindet, dieselbe Bedeutung haben, da sich dies aus den beiden Ausdrücken selbst ersehen läßt.

6.2321 Und, daß die Sätze der Mathematik bewiesen werden können, heißt ja nichts anderes, als daß ihre Richtigkeit

6.13 Logic is not a body of doctrine, but a mirror-image of the world.

Logic is transcendental.

6.2 Mathematics is a logical method.

The propositions of mathematics are equations, and therefore pseudo-propositions.

6.21 A proposition of mathematics does not express a thought.

6.211 Indeed in real life a mathematical proposition is never what we want. Rather, we make use of mathematical propositions *only* in inferences from propositions that do not belong to mathematics to others that likewise do not belong to mathematics.

(In philosophy the question, 'What do we actually use this word or this proposition for?' repeatedly leads to valuable insights.)

6.22 The logic of the world, which is shown in tautologies by the propositions of logic, is shown in equations by mathematics.

6.23 If two expressions are combined by means of the sign of equality, that means that they can be substituted for one another. But it must be manifest in the two expressions themselves whether this is the case or not.

When two expressions can be substituted for one another, that characterizes their logical form.

6.231 It is a property of affirmation that it can be construed as double negation.

It is a property of '$1+1+1+1$' that it can be construed as '$(1+1)+(1+1)$'.

6.232 Frege says that the two expressions have the same meaning but different senses.

But the essential point about an equation is that it is not necessary in order to show that the two expressions connected by the sign of equality have the same meaning, since this can be seen from the two expressions themselves.

6.2321 And the possibility of proving the propositions of mathematics means simply that their correctness can be

einzusehen ist, ohne daß das, was sie ausdrücken, selbst mit den Tatsachen auf seine Richtigkeit hin verglichen werden muß.

6.2322 Die Identität der Bedeutung zweier Ausdrücke läßt sich nicht *behaupten*. Denn, um etwas von ihrer Bedeutung behaupten zu können, muß ich ihre Bedeutung kennen: und indem ich ihre Bedeutung kenne, weiß ich, ob sie dasselbe oder verschiedenes bedeuten.

6.2323 Die Gleichung kennzeichnet nur den Standpunkt, von welchem ich die beiden Ausdrücke betrachte, nämlich vom Standpunkte ihrer Bedeutungsgleichheit.

6.233 Die Frage, ob man zur Lösung der mathematischen Probleme die Anschauung brauche, muß dahin beantwortet werden, daß eben die Sprache hier die nötige Anschauung liefert.

6.2331 Der Vorgang des *Rechnens* vermittelt eben diese Anschauung.

Die Rechnung ist kein Experiment.

6.234 Die Mathematik ist eine Methode der Logik.

6.2341 Das Wesentliche der mathematischen Methode ist es, mit Gleichungen zu arbeiten. Auf dieser Methode beruht es nämlich, daß jeder Satz der Mathematik sich von selbst verstehen muß.

6.24 Die Methode der Mathematik, zu ihren Gleichungen zu kommen, ist die Substitutionsmethode.

Denn die Gleichungen drücken die Ersetzbarkeit zweier Ausdrücke aus und wir schreiten von einer Anzahl von Gleichungen zu neuen Gleichungen vor, indem wir, den Gleichungen entsprechend, Ausdrücke durch andere ersetzen.

6.241 So lautet der Beweis des Satzes $2 \times 2 = 4$:

$$(\Omega^{\nu})^{\mu}{}'x = \Omega^{\nu \times \mu}{}'x \text{ Def.},$$

$$\begin{aligned}\Omega^{2 \times 2}{}'x &= (\Omega^{2})^{2}{}'x = (\Omega^{2})^{1+1}{}'x \\ &= \Omega^{2}{}'\Omega^{2}{}'x = \Omega^{1+1}{}'\Omega^{1+1}{}'x = (\Omega'\Omega)'(\Omega'\Omega)'x \\ &= \Omega'\Omega'\Omega'\Omega'x = \Omega^{1+1+1+1}{}'x = \Omega^{4}{}'x.\end{aligned}$$

perceived without its being necessary that what they express should itself be compared with the facts in order to determine its correctness.

6.2322 It is impossible to *assert* the identity of meaning of two expressions. For in order to be able to assert anything about their meaning, I must know their meaning, and I cannot know their meaning without knowing whether what they mean is the same or different.

6.2323 An equation merely marks the point of view from which I consider the two expressions: it marks their equivalence in meaning.

6.233 The question whether intuition is needed for the solution of mathematical problems must be given the answer that in this case language itself provides the necessary intuition.

6.2331 The process of *calculating* serves to bring about that intuition.

Calculation is not an experiment.

6.234 Mathematics is a method of logic.

6.2341 It is the essential characteristic of mathematical method that it employs equations. For it is because of this method that every proposition of mathematics must go without saying.

6.24 The method by which mathematics arrives at its equations is the method of substitution.

For equations express the substitutability of two expressions and, starting from a number of equations, we advance to new equations by substituting different expressions in accordance with the equations.

6.241 Thus the proof of the proposition $2 \times 2 = 4$ runs as follows:

$$(\Omega^{\nu})^{\mu\text{'}}x = \Omega^{\nu \times \mu\text{'}}x \text{ Def.,}$$

$$\begin{aligned}\Omega^{2 \times 2\text{'}}x &= (\Omega^{2})^{2\text{'}}x = (\Omega^{2})^{1+1\text{'}}x \\ &= \Omega^{2\text{'}}\Omega^{2\text{'}}x = \Omega^{1+1\text{'}}\Omega^{1+1\text{'}}x = (\Omega\text{'}\Omega)\text{'}(\Omega\text{'}\Omega)\text{'}x \\ &= \Omega\text{'}\Omega\text{'}\Omega\text{'}\Omega\text{'}x = \Omega^{1+1+1+1\text{'}}x = \Omega^{4\text{'}}x.\end{aligned}$$

6.3 Die Erforschung der Logik bedeutet die Erforschung *aller Gesetzmäßigkeit*. Und außerhalb der Logik ist alles Zufall.

6.31 Das sogenannte Gesetz der Induktion kann jedenfalls kein logisches Gesetz sein, denn es ist offenbar ein sinnvoller Satz.— Und darum kann es auch kein Gesetz a priori sein.

6.32 Das Kausalitätsgesetz ist kein Gesetz, sondern die Form eines Gesetzes.

6.321 „Kausalitätsgesetz", das ist ein Gattungsname. Und wie es in der Mechanik, sagen wir, Minimum-Gesetze gibt — etwa der kleinsten Wirkung —, so gibt es in der Physik Kausalitätsgesetze, Gesetze von der Kausalitätsform.

6.3211 Man hat ja auch davon eine Ahnung gehabt, daß es *ein* „Gesetz der kleinsten Wirkung" geben müsse, ehe man genau wußte, wie es lautete. (Hier, wie immer, stellt sich das a priori Gewisse als etwas rein Logisches heraus.)

6.33 Wir *glauben* nicht a priori an ein Erhaltungsgesetz, sondern wir *wissen* a priori die Möglichkeit einer logischen Form.

6.34 Alle jene Sätze, wie der Satz vom Grunde, von der Kontinuität in der Natur, vom kleinsten Aufwande in der Natur, etc. etc., alle diese sind Einsichten a priori über die mögliche Formgebung der Sätze der Wissenschaft.

6.341 Die Newtonsche Mechanik z. B. bringt die Weltbeschreibung auf eine einheitliche Form. Denken wir uns eine weiße Fläche, auf der unregelmäßige schwarze Flecken wären. Wir sagen nun: Was für ein Bild immer hierdurch entsteht, immer kann ich seiner Beschreibung beliebig nahe kommen, indem ich die Fläche mit einem entsprechend feinen quadratischen Netzwerk bedecke und nun von jedem Quadrat sage, daß es weiß oder schwarz ist. Ich werde auf diese Weise die Beschreibung der Fläche auf eine einheitliche Form gebracht haben. Diese Form ist beliebig, denn ich hätte mit dem gleichen Erfolge ein Netz aus dreieckigen oder sechseckigen Maschen verwenden können. Es kann sein, daß die Beschreibung mit Hilfe eines Dreiecks-Netzes einfacher geworden wäre; das heißt,

6.3 The exploration of logic means the exploration of *everything that is subject to law*. And outside logic everything is accidental.

6.31 The so-called law of induction cannot possibly be a law of logic, since it is obviously a proposition with sense.—Nor, therefore, can it be an a priori law.

6.32 The law of causality is not a law but the form of a law.

6.321 'Law of causality'—that is a general name. And just as in mechanics, for example, there are 'minimum-principles', such as the law of least action, so too in physics there are causal laws, laws of the causal form.

6.3211 Indeed people even surmised that there must be *a* 'law of least action' before they knew exactly how it went. (Here, as always, what is certain a priori proves to be something purely logical.)

6.33 We do not have an a priori *belief* in a law of conservation, but rather a priori *knowledge* of the possibility of a logical form.

6.34 All such propositions, including the principle of sufficient reason, the laws of continuity in nature and of least effort in nature, etc. etc.—all these are a priori insights about the forms in which the propositions of science can be cast.

6.341 Newtonian mechanics, for example, imposes a unified form on the description of the world. Let us imagine a white surface with irregular black spots on it. We then say that whatever kind of picture these make, I can always approximate as closely as I wish to the description of it by covering the surface with a sufficiently fine square mesh, and then saying of every square whether it is black or white. In this way I shall have imposed a unified form on the description of the surface. The form is optional, since I could have achieved the same result by using a net with a triangular or hexagonal mesh. Possibly the use of a triangular mesh would have made the description simpler: that is to say, it might be that we could describe the surface more accurately with a coarse triangular mesh than

daß wir die Fläche mit einem gröberen Dreiecks-Netz genauer beschreiben könnten als mit einem feineren quadratischen (oder umgekehrt), usw. Den verschiedenen Netzen entsprechen verschiedene Systeme der Weltbeschreibung. Die Mechanik bestimmt eine Form der Weltbeschreibung, indem sie sagt: Alle Sätze der Weltbeschreibung müssen aus einer Anzahl gegebener Sätze — den mechanischen Axiomen — auf eine gegebene Art und Weise erhalten werden. Hierdurch liefert sie die Bausteine zum Bau des wissenschaftlichen Gebäudes und sagt: Welches Gebäude immer du aufführen willst, jedes mußt du irgendwie mit diesen und nur diesen Bausteinen zusammenbringen.

(Wie man mit dem Zahlensystem jede beliebige Anzahl, so muß man mit dem System der Mechanik jeden beliebigen Satz der Physik hinschreiben können.)

6.342 Und nun sehen wir die gegenseitige Stellung von Logik und Mechanik. (Man könnte das Netz auch aus verschiedenartigen Figuren etwa aus Dreiecken und Sechsecken bestehen lassen.) Daß sich ein Bild, wie das vorhin erwähnte, durch ein Netz von gegebener Form beschreiben läßt, sagt über das Bild *nichts* aus. (Denn dies gilt für jedes Bild dieser Art.) *Das* aber charakterisiert das Bild, daß es sich durch ein bestimmtes Netz von *bestimmter* Feinheit *vollständig* beschreiben läßt.

So auch sagt es nichts über die Welt aus, daß sie sich durch die Newtonsche Mechanik beschreiben läßt; wohl aber, daß sie sich *so* durch jene beschreiben läßt, wie dies eben der Fall ist. Auch das sagt etwas über die Welt, daß sie sich durch die eine Mechanik einfacher beschreiben läßt als durch die andere.

6.343 Die Mechanik ist ein Versuch, alle *wahren* Sätze, die wir zur Weltbeschreibung brauchen, nach Einem Plane zu konstruieren.

6.3431 Durch den ganzen logischen Apparat hindurch sprechen die physikalischen Gesetze doch von den Gegenständen der Welt.

6.3432 Wir dürfen nicht vergessen, daß die Weltbeschreibung

with a fine square mesh (or conversely), and so on. The different nets correspond to different systems for describing the world. Mechanics determines one form of description of the world by saying that all propositions used in the description of the world must be obtained in a given way from a given set of propositions—the axioms of mechanics. It thus supplies the bricks for building the edifice of science, and it says, 'Any building that you want to erect, whatever it may be, must somehow be constructed with these bricks, and with these alone.'

(Just as with the number-system we must be able to write down any number we wish, so with the system of mechanics we must be able to write down any proposition of physics that we wish.)

6.342 And now we can see the relative position of logic and mechanics. (The net might also consist of more than one kind of mesh: e.g. we could use both triangles and hexagons.) The possibility of describing a picture like the one mentioned above with a net of a given form tells us *nothing* about the picture. (For that is true of all such pictures.) But what *does* characterize the picture is that it can be described *completely* by a particular net with a *particular* size of mesh.

Similarly the possibility of describing the world by means of Newtonian mechanics tells us nothing about the world: but what does tell us something about it is the precise *way* in which it is possible to describe it by these means. We are also told something about the world by the fact that it can be described more simply with one system of mechanics than with another.

6.343 Mechanics is an attempt to construct according to a single plan all the *true* propositions that we need for the description of the world.

6.3431 The laws of physics, with all their logical apparatus, still speak, however indirectly, about the objects of the world.

6.3432 We ought not to forget that any description of the

durch die Mechanik immer die ganz allgemeine ist. Es ist in ihr z. B. nie von *bestimmten* materiellen Punkten die Rede, sondern immer nur von *irgend welchen.*

6.35 Obwohl die Flecke in unserem Bild geometrische Figuren sind, so kann doch selbstverständlich die Geometrie gar nichts über ihre tatsächliche Form und Lage sagen. Das Netz aber ist *rein* geometrisch, alle seine Eigenschaften können a priori angegeben werden.

Gesetze wie der Satz vom Grunde, etc. handeln vom Netz, nicht von dem, was das Netz beschreibt.

6.36 Wenn es ein Kausalitätsgesetz gäbe, so könnte es lauten: „Es gibt Naturgesetze".

Aber freilich kann man das nicht sagen: es zeigt sich.

6.361 In der Ausdrucksweise Hertz' könnte man sagen: Nur *gesetzmäßige* Zusammenhänge sind *denkbar.*

6.3611 Wir können keinen Vorgang mit dem „Ablauf der Zeit" vergleichen — diesen gibt es nicht —, sondern nur mit einem anderen Vorgang (etwa mit dem Gang des Chronometers).

Daher ist die Beschreibung des zeitlichen Verlaufs nur so möglich, daß wir uns auf einen anderen Vorgang stützen.

Ganz Analoges gilt für den Raum. Wo man z. B. sagt, es könne keines von zwei Ereignissen (die sich gegenseitig ausschließen) eintreten, weil *keine Ursache* vorhanden sei, warum das eine eher als das andere eintreten solle, da handelt es sich in Wirklichkeit darum, daß man gar nicht *eines* der beiden Ereignisse beschreiben kann, wenn nicht irgend eine Asymmetrie vorhanden ist. Und *wenn* eine solche Asymmetrie vorhanden *ist*, so können wir diese als *Ursache* des Eintreffens des einen und Nicht-Eintreffens des anderen auffassen.

6.36111 Das Kantsche Problem von der rechten und linken Hand, die man nicht zur Deckung bringen kann, besteht schon in der Ebene, ja im eindimensionalen Raum,

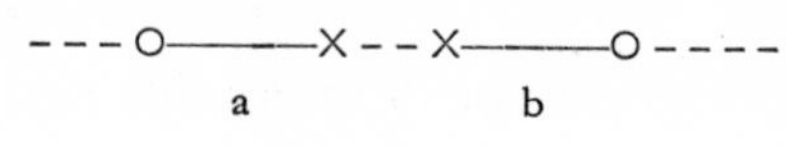

world by means of mechanics will be of the completely general kind. For example, it will never mention *particular* point-masses: it will only talk about *any point-masses whatsoever*.

6.35 Although the spots in our picture are geometrical figures, nevertheless geometry can obviously say nothing at all about their actual form and position. The network, however, is *purely* geometrical; all its properties can be given a priori.

Laws like the principle of sufficient reason, etc. are about the net and not about what the net describes.

6.36 If there were a law of causality, it might be put in the following way: There are laws of nature.

But of course that cannot be said: it makes itself manifest.

6.361 One might say, using Hertz's terminology, that only connexions that are *subject to law* are *thinkable*.

6.3611 We cannot compare a process with 'the passage of time'—there is no such thing—but only with another process (such as the working of a chronometer).

Hence we can describe the lapse of time only by relying on some other process.

Something exactly analogous applies to space: e.g. when people say that neither of two events (which exclude one another) can occur, because there is *nothing to cause* the one to occur rather than the other, it is really a matter of our being unable to describe *one* of the two events unless there is some sort of asymmetry to be found. And *if* such an asymmetry *is* to be found, we can regard it as the *cause* of the occurrence of the one and the non-occurrence of the other.

6.36111 Kant's problem about the right hand and the left hand, which cannot be made to coincide, exists even in two dimensions. Indeed, it exists in one-dimensional space

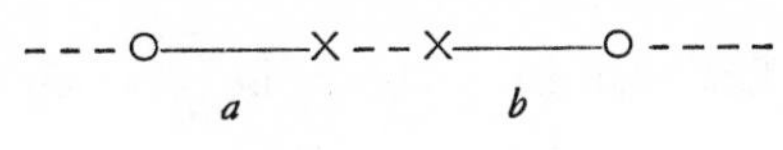

wo die beiden kongruenten Figuren a und b auch nicht zur Deckung gebracht werden können, ohne aus diesem Raum herausbewegt zu werden. Rechte und linke Hand sind tatsächlich vollkommen kongruent. Und daß man sie nicht zur Deckung bringen kann, hat damit nichts zu tun.

Den rechten Handschuh könnte man an die linke Hand ziehen, wenn man ihn im vierdimensionalen Raum umdrehen könnte.

6.362 Was sich beschreiben läßt, das kann auch geschehen, und was das Kausalitätsgesetz ausschließen soll, das läßt sich auch nicht beschreiben.

6.363 Der Vorgang der Induktion besteht darin, daß wir das *einfachste* Gesetz annehmen, das mit unseren Erfahrungen in Einklang zu bringen ist.

6.3631 Dieser Vorgang hat aber keine logische, sondern nur eine psychologische Begründung.

Es ist klar, daß kein Grund vorhanden ist, zu glauben, es werde nun auch wirklich der einfachste Fall eintreten.

6.36311 Daß die Sonne morgen aufgehen wird, ist eine Hypothese; und das heißt: Wir *wissen* nicht, ob sie aufgehen wird.

6.37 Einen Zwang, nach dem Eines geschehen müßte, weil etwas anderes geschehen ist, gibt es nicht. Es gibt nur eine *logische* Notwendigkeit.

6.371 Der ganzen modernen Weltanschauung liegt die Täuschung zugrunde, daß die sogenannten Naturgesetze die Erklärungen der Naturerscheinungen seien.

6.372 So bleiben sie bei den Naturgesetzen als bei etwas Unantastbarem stehen, wie die Älteren bei Gott und dem Schicksal.

Und sie haben ja beide Recht, und Unrecht. Die Alten sind allerdings insofern klarer, als sie einen klaren Abschluß anerkennen, während es bei dem neuen System scheinen soll, als sei *alles* erklärt.

6.373 Die Welt ist unabhängig von meinem Willen.

in which the two congruent figures, *a* and *b*, cannot be made to coincide unless they are moved out of this space. The right hand and the left hand are in fact completely congruent. It is quite irrelevant that they cannot be made to coincide.

A right-hand glove could be put on the left hand, if it could be turned round in four-dimensional space.

6.362 What can be described can happen too: and what the law of causality is meant to exclude cannot even be described.

6.363 The procedure of induction consists in accepting as true the *simplest* law that can be reconciled with our experiences.

6.3631 This procedure, however, has no logical justification but only a psychological one.

It is clear that there are no grounds for believing that the simplest eventuality will in fact be realized.

6.36311 It is an hypothesis that the sun will rise tomorrow: and this means that we do not *know* whether it will rise.

6.37 There is no compulsion making one thing happen because another has happened. The only necessity that exists is *logical* necessity.

6.371 The whole modern conception of the world is founded on the illusion that the so-called laws of nature are the explanations of natural phenomena.

6.372 Thus people today stop at the laws of nature, treating them as something inviolable, just as God and Fate were treated in past ages.

And in fact both are right and both wrong: though the view of the ancients is clearer in so far as they have a clear and acknowledged terminus, while the modern system tries to make it look as if *everything* were explained.

6.373 The world is independent of my will.

6.374 Auch wenn alles, was wir wünschen, geschähe, so wäre dies doch nur, sozusagen, eine Gnade des Schicksals, denn es ist kein logischer Zusammenhang zwischen Willen und Welt, der dies verbürgte, und den angenommenen physikalischen Zusammenhang könnten wir doch nicht selbst wieder wollen.

6.375 Wie es nur eine logische Notwendigkeit gibt, so gibt es auch nur eine logische Unmöglichkeit.

6.3751 Daß z. B. zwei Farben zugleich an einem Ort des Gesichtsfeldes sind, ist unmöglich, und zwar logisch unmöglich, denn es ist durch die logische Struktur der Farbe ausgeschlossen.

Denken wir daran, wie sich dieser Widerspruch in der Physik darstellt: Ungefähr so, daß ein Teilchen nicht zu gleicher Zeit zwei Geschwindigkeiten haben kann; das heißt, daß es nicht zu gleicher Zeit an zwei Orten sein kann; das heißt, daß Teilchen an verschiedenen Orten zu Einer Zeit nicht identisch sein können.

(Es ist klar, daß das logische Produkt zweier Elementarsätze weder eine Tautologie noch eine Kontradiktion sein kann. Die Aussage, daß ein Punkt des Gesichtsfeldes zu gleicher Zeit zwei verschiedene Farben hat, ist eine Kontradiktion.)

6.4 Alle Sätze sind gleichwertig.

6.41 Der Sinn der Welt muß außerhalb ihrer liegen. In der Welt ist alles, wie es ist, und geschieht alles, wie es geschieht; es gibt in ihr keinen Wert — und wenn es ihn gäbe, so hätte er keinen Wert.

Wenn es einen Wert gibt, der Wert hat, so muß er außerhalb alles Geschehens und So-Seins liegen. Denn alles Geschehen und So-Sein ist zufällig.

Was es nichtzufällig macht, kann nicht in der Welt liegen, denn sonst wäre dies wieder zufällig.

Es muß außerhalb der Welt liegen.

6.42 Darum kann es auch keine Sätze der Ethik geben.

Sätze können nichts Höheres ausdrücken.

6.374 Even if all that we wish for were to happen, still this would only be a favour granted by fate, so to speak: for there is no *logical* connexion between the will and the world, which would guarantee it, and the supposed physical connexion itself is surely not something that we could will.

6.375 Just as the only necessity that exists is *logical* necessity, so too the only impossibility that exists is *logical* impossibility.

6.3751 For example, the simultaneous presence of two colours at the same place in the visual field is impossible, in fact logically impossible, since it is ruled out by the logical structure of colour.

Let us think how this contradiction appears in physics: more or less as follows—a particle cannot have two velocities at the same time; that is to say, it cannot be in two places at the same time; that is to say, particles that are in different places at the same time cannot be identical.

(It is clear that the logical product of two elementary propositions can neither be a tautology nor a contradiction. The statement that a point in the visual field has two different colours at the same time is a contradiction.)

6.4 All propositions are of equal value.

6.41 The sense of the world must lie outside the world. In the world everything is as it is, and everything happens as it does happen: *in* it no value exists—and if it did exist, it would have no value.

If there is any value that does have value, it must lie outside the whole sphere of what happens and is the case. For all that happens and is the case is accidental.

What makes it non-accidental cannot lie *within* the world, since if it did it would itself be accidental.

It must lie outside the world.

6.42 So too it is impossible for there to be propositions of ethics.

Propositions can express nothing that is higher.

6.421 Es ist klar, daß sich die Ethik nicht aussprechen läßt.

Die Ethik ist transzendental.

(Ethik und Ästhetik sind Eins.)

6.422 Der erste Gedanke bei der Aufstellung eines ethischen Gesetzes von der Form „Du sollst . . . .“ ist: Und was dann, wenn ich es nicht tue? Es ist aber klar, daß die Ethik nichts mit Strafe und Lohn im gewöhnlichen Sinne zu tun hat. Also muß diese Frage nach den *Folgen* einer Handlung belanglos sein.— Zum Mindesten dürfen diese Folgen nicht Ereignisse sein. Denn etwas muß doch an jener Fragestellung richtig sein. Es muß zwar eine Art von ethischem Lohn und ethischer Strafe geben, aber diese müssen in der Handlung selbst liegen.

(Und das ist auch klar, daß der Lohn etwas Angenehmes, die Strafe etwas Unangenehmes sein muß.)

6.423 Vom Willen als dem Träger des Ethischen kann nicht gesprochen werden.

Und der Wille als Phänomen interessiert nur die Psychologie.

6.43 Wenn das gute oder böse Wollen die Welt ändert, so kann es nur die Grenzen der Welt ändern, nicht die Tatsachen; nicht das, was durch die Sprache ausgedrückt werden kann.

Kurz, die Welt muß dann dadurch überhaupt eine andere werden. Sie muß sozusagen als Ganzes abnehmen oder zunehmen.

Die Welt des Glücklichen ist eine andere als die des Unglücklichen.

6.431 Wie auch beim Tod die Welt sich nicht ändert, sondern aufhört.

6.4311 Der Tod ist kein Ereignis des Lebens. Den Tod erlebt man nicht.

Wenn man unter Ewigkeit nicht unendliche Zeitdauer, sondern Unzeitlichkeit versteht, dann lebt der ewig, der in der Gegenwart lebt.

Unser Leben ist ebenso endlos, wie unser Gesichtsfeld grenzenlos ist.

6.4312 Die zeitliche Unsterblichkeit der Seele des Menschen, das heißt also ihr ewiges Fortleben auch nach dem Tode,

6.421 It is clear that ethics cannot be put into words.

Ethics is transcendental.

(Ethics and aesthetics are one and the same.)

6.422 When an ethical law of the form, 'Thou shalt . . .', is laid down, one's first thought is, 'And what if I do not do it?' It is clear, however, that ethics has nothing to do with punishment and reward in the usual sense of the terms. So our question about the *consequences* of an action must be unimportant.—At least those consequences should not be events. For there must be something right about the question we posed. There must indeed be some kind of ethical reward and ethical punishment, but they must reside in the action itself.

(And it is also clear that the reward must be something pleasant and the punishment something unpleasant.)

6.423 It is impossible to speak about the will in so far as it is the subject of ethical attributes.

And the will as a phenomenon is of interest only to psychology.

6.43 If the good or bad exercise of the will does alter the world, it can alter only the limits of the world, not the facts—not what can be expressed by means of language.

In short the effect must be that it becomes an altogether different world. It must, so to speak, wax and wane as a whole.

The world of the happy man is a different one from that of the unhappy man.

6.431 So too at death the world does not alter, but comes to an end.

6.4311 Death is not an event in life: we do not live to experience death.

If we take eternity to mean not infinite temporal duration but timelessness, then eternal life belongs to those who live in the present.

Our life has no end in just the way in which our visual field has no limits.

6.4312 Not only is there no guarantee of the temporal immortality of the human soul, that is to say of its eternal sur-

ist nicht nur auf keine Weise verbürgt, sondern vor allem leistet diese Annahme gar nicht das, was man immer mit ihr erreichen wollte. Wird denn dadurch ein Rätsel gelöst, daß ich ewig fortlebe? Ist denn dieses ewige Leben dann nicht ebenso rätselhaft wie das gegenwärtige? Die Lösung des Rätsels des Lebens in Raum und Zeit liegt *außerhalb* von Raum und Zeit.

(Nicht Probleme der Naturwissenschaft sind ja zu lösen.)

6.432 *Wie* die Welt ist, ist für das Höhere vollkommen gleichgültig. Gott offenbart sich nicht *in* der Welt.

6.4321 Die Tatsachen gehören alle nur zur Aufgabe, nicht zur Lösung.

6.44 Nicht *wie* die Welt ist, ist das Mystische, sondern *daß* sie ist.

6.45 Die Anschauung der Welt sub specie aeterni ist ihre Anschauung als — begrenztes — Ganzes.

Das Gefühl der Welt als begrenztes Ganzes ist das mystische.

6.5 Zu einer Antwort, die man nicht aussprechen kann, kann man auch die Frage nicht aussprechen.

Das Rätsel gibt es nicht.

Wenn sich eine Frage überhaupt stellen läßt, so *kann* sie auch beantwortet werden.

6.51 Skeptizismus ist *nicht* unwiderleglich, sondern offenbar unsinnig, wenn er bezweifeln will, wo nicht gefragt werden kann.

Denn Zweifel kann nur bestehen, wo eine Frage besteht; eine Frage nur, wo eine Antwort besteht, und diese nur, wo etwas *gesagt* werden *kann*.

6.52 Wir fühlen, daß, selbst wenn alle möglichen wissenschaftlichen Fragen beantwortet sind, unsere Lebensprobleme noch gar nicht berührt sind. Freilich bleibt dann eben keine Frage mehr; und eben dies ist die Antwort.

6.521 Die Lösung des Problems des Lebens merkt man am Verschwinden dieses Problems.

(Ist nicht dies der Grund, warum Menschen, denen

vival after death; but, in any case, this assumption completely fails to accomplish the purpose for which it has always been intended. Or is some riddle solved by my surviving for ever? Is not this eternal life itself as much of a riddle as our present life? The solution of the riddle of life in space and time lies *outside* space and time.

(It is certainly not the solution of any problems of natural science that is required.)

6.432 *How* things are in the world is a matter of complete indifference for what is higher. God does not reveal himself *in* the world.

6.4321 The facts all contribute only to setting the problem, not to its solution.

6.44 It is not *how* things are in the world that is mystical, but *that* it exists.

6.45 To view the world sub specie aeterni is to view it as a whole—a limited whole.

Feeling the world as a limited whole—it is this that is mystical.

6.5 When the answer cannot be put into words, neither can the question be put into words.

The riddle does not exist.

If a question can be framed at all, it is also *possible* to answer it.

6.51 Scepticism is *not* irrefutable, but obviously nonsensical, when it tries to raise doubts where no questions can be asked.

For doubt can exist only where a question exists, a question only where an answer exists, and an answer only where something *can be said*.

6.52 We feel that even when all *possible* scientific questions have been answered, the problems of life remain completely untouched. Of course there are then no questions left, and this itself is the answer.

6.521 The solution of the problem of life is seen in the vanishing of the problem.

(Is not this the reason why those who have found after

der Sinn des Lebens nach langen Zweifeln klar wurde, warum diese dann nicht sagen konnten, worin dieser Sinn bestand?)

6.522 Es gibt allerdings Unaussprechliches. Dies z e i g t sich, es ist das Mystische.

6.53 Die richtige Methode der Philosophie wäre eigentlich die: Nichts zu sagen, als was sich sagen läßt, also Sätze der Naturwissenschaft — also etwas, was mit Philosophie nichts zu tun hat —, und dann immer, wenn ein anderer etwas Metaphysisches sagen wollte, ihm nachzuweisen, daß er gewissen Zeichen in seinen Sätzen keine Bedeutung gegeben hat. Diese Methode wäre für den anderen unbefriedigend — er hätte nicht das Gefühl, daß wir ihn Philosophie lehrten — aber s i e wäre die einzig streng richtige.

6.54 Meine Sätze erläutern dadurch, daß sie der, welcher mich versteht, am Ende als unsinnig erkennt, wenn er durch sie — auf ihnen — über sie hinausgestiegen ist. (Er muß sozusagen die Leiter wegwerfen, nachdem er auf ihr hinaufgestiegen ist.)

Er muß diese Sätze überwinden, dann sieht er die Welt richtig.

7 Wovon man nicht sprechen kann, darüber muß man schweigen.

a long period of doubt that the sense of life became clear to them have then been unable to say what constituted that sense?)

6.522 There are, indeed, things that cannot be put into words. They *make themselves manifest*. They are what is mystical.

6.53 The correct method in philosophy would really be the following: to say nothing except what can be said, i.e. propositions of natural science—i.e. something that has nothing to do with philosophy—and then, whenever someone else wanted to say something metaphysical, to demonstrate to him that he had failed to give a meaning to certain signs in his propositions. Although it would not be satisfying to the other person—he would not have the feeling that we were teaching him philosophy—*this* method would be the only strictly correct one.

6.54 My propositions serve as elucidations in the following way: anyone who understands me eventually recognizes them as nonsensical, when he has used them—as steps—to climb up beyond them. (He must, so to speak, throw away the ladder after he has climbed up it.)

He must transcend these propositions, and then he will see the world aright.

7 What we cannot speak about we must pass over in silence.

INDEX

The translators' aim has been to include all the more interesting words, and, in each case, either to give all the occurrences of a word, or else to omit only a few unimportant ones. Paragraphs in the preface are referred to as P1, P2, etc. Propositions are indicated by numbers without points; more than two consecutive propositions, by two numbers joined by an en-rule, as 202–2021.

In the translation it has sometimes been necessary to use different English expressions for the same German expression or the same English expression for different German expressions. The index contains various devices designed to make it an informative guide to the German terminology and, in particular, to draw attention to some important connexions between ideas that are more difficult to bring out in English than in German.

First, when a German expression is of any interest in itself, it is given in brackets after the English expression that translates it, e.g. **situation** [*Sachlage*]; also, whenever an English expression is used to translate more than one German expression, each of the German expressions is given separately in numbered brackets, and is followed by the list of passages in which it is translated by the English expression, e.g. **reality 1.** [*Realität*], 55561, etc. **2.** [*Wirklichkeit*], 206, etc.

Secondly, the German expressions given in this way sometimes have two or more English translations in the text; and when this is so, if the alternative English translations are of interest, they follow the German expression inside the brackets, e.g. **proposition** [*Satz*: law; principle].

The alternative translations recorded by these two devices are sometimes given in an abbreviated way. For a German expression need not actually be translated by the English expressions that it follows or precedes, as it is in the examples above. The relationship may be more complicated. For instance, the German expression may be only part of a phrase that is translated by the English expression, e.g. **stand in a relation to one another; are related** [*sich verhalten*: stand, how things; state of things].

Thirdly, cross-references have been used to draw attention to other important connexions between ideas, e.g. **true,** cf. correct; right: and ***a priori,*** cf. advance, in.

In subordinate entries and cross-references the catchword is indicated by ~, unless the catchword contains /, in which case the part preceding / is so indicated, e.g. **accident**; **~al** for **accident**; **accidental,** and **state of / affairs**; **~ things** for **state of affairs**; **state of things.** Cross-references relate to the last preceding entry or numbered bracket. When references are given both for a word in its own right and for a phrase containing it, occurrences of the latter are generally not also counted as occurrences of the former, so that both entries should be consulted.